21ST CENTURY LEADERSHIP

TO **FIGHT THE CODE RED**
FOR BUSINESS

R. A. FERNANDO

ARCHWAY
PUBLISHING

Archway Publishing books may be ordered through booksellers or by contacting:

Archway Publishing
1663 Liberty Drive
Bloomington, IN 47403
www.archwaypublishing.com
844-669-3957

ISBN: 978-1-6657-3565-0 (sc)
ISBN: 978-1-6657-3564-3 (hc)
ISBN: 978-1-6657-3563-6 (e)

Library of Congress Control Number: 2022923977

Print information available on the last page.

Archway Publishing rev. date: 02/06/2023

Global Strategic Corporate Sustainability Pvt. Ltd. was incorporated on February 10, 2015. The directors are Dr. R. A. Fernando and Charith Fernando. Following are US copyrights and intellectual property rights they own:

- Strategic Corporate Sustainability©, US copyright TX 7-180-253, January 31, 2010.
- *Strategic Corporate Sustainability: 7 Imperatives for Sustainable Business* (Partridge, 2015), which is based on Dr. R. A. Fernando's work at Cambridge University (2007–2014). ISBN 978-1-4828-5404-6.
- Twenty-First-Century Board Leadership Model©, US copyright TX 8-880-097, June 17, 2020, which was the inspiration and basis for Dr. Fernando's second book, *Twenty-First-Century Leadership to Fight the Code Red for Business.*
- Twenty-First-Century Board Leadership Model MasterClass© created jointly by Dr. R. A. Fernando and Raymond Schadeck, which was launched in Luxembourg in September 2020 by the Institut Luxembourgeois des Administrateurs (ILA).
- *Twenty-First-Century Leadership to Fight the Code Red for Business*, which is based on the Twenty-First-Century Board Leadership Model created by Dr. R. A. Fernando in 2020.

Dedication

To my mother, Nalini, who encouraged me with prayer every day of her life; my wife, Varuni; my son Charith; and my sister Sureshmie, all of whom supported me and believed that world-class quality and global impact are both possible to achieve if one is focused and passionate.

CONTENTS

LIST OF ILLUSTRATIONS

ABOUT THE AUTHOR

Dr. R. A. Fernando is an alumnus of the University of Cambridge, having completed a postgraduate certificate in sustainable business in 2008

and a master of studies in sustainability leadership in 2014. He holds a doctor of business administration degree from the European Business School, which he earned in 2016. He has an MBA from the University of Colombo and is a fellow of the Chartered Institute of Marketing (UK).

At the INSEAD Business School (France), he completed both a diploma in international management and the Advanced Management Program. He has been an executive in residence at the INSEAD Business School since 2010 and is a visiting faculty member of the INSEAD Advanced Strategy for Directors program and the Aspiring Directors program.

In 2010, Dr. Fernando received a US copyright for his work at Cambridge University for the concept of strategic corporate sustainability. In November 2015, he published *Strategic Corporate Sustainability: 7 Imperatives for Sustainable Business* (Partridge), based on his work at Cambridge University. In 2017, he created the ILA Future Directors program for the Institute Luxembourgeois des Administrateurs, which set the foundation for the Twenty-First-Century Board Leadership Model, which was created on June 17, 2020, and the Twenty-First-Century

Board Leadership Model MasterClass©, which was launched by ILA in Luxembourg in September 2020. Today, the MasterClass is building a new cadre of twenty-first-century leaders in Luxembourg, Ireland, and Sri Lanka and is to be expanded to many nations as a key supplement to the existing knowledge of those people in positions of leadership who want to be relevant twenty-first-century leaders. The book *Twenty-First-Century Leadership to Fight the Code Red for Business* is the culmination of this journey to build future-ready twenty-first-century leaders.

Dr. Fernando is the chairman and CEO of Global Strategic Corporate Sustainability Pvt. Ltd., His career with multinationals spanned 1981–2007 with Unilever, Reckitt Benckiser, and SmithKline Beecham International, covering Africa, the Middle East, and Asia in CEO / business development positions. He was the first CEO of the Sri Lanka Institute of Nanotechnology, 2008–2011, and operations director of the Malaysia Blue Ocean Strategy Institute, 2011–2016. He was the UN Global Compact focal point in 2007–2011 and set up the UNGC Sri Lanka Network.

Dr. Fernando serves on the boards of Global Strategic Corporate Sustainability Pvt. Ltd., Dilmah Ceylon Tea PLC, Aitken Spence Plantations Ltd., LOLC Holdings, Ceylon Graphene Technologies Ltd., Habitat for Humanity, UN Global Compact, and Ceylon Asset Management Ltd. In September 2007 he won the Global Strategy Leadership Award, presented to him by Professor Renee Mauborgne of INSEAD at the World Strategy summit in Mumbai. In February 2020, he won the World CSR Leadership Award.

Dr. Fernando is deeply committed to and passionate about creating a cadre of twenty-first-century, science-led, sustainability-mindset leaders to make an impact on the code red for humanity and business.

FOREWORDS

I first met Dr. R. A. Fernando in 2016, at a conference hosted by the INSEAD Alumni Association of Luxembourg and ILA (Institut Luxembourgeois des Administrateurs), which I had the pleasure of chairing at the time. Since this first crossing, Ravi and I seem to have been bonded together as we've worked to organize numerous conferences and training programs.

It's a bond based on a number of shared beliefs that Ravi so eloquently brings to the surface.

Since our first meeting, I've been fascinated by Ravi's visionary thinking in framing our future less like a desert of hopeless challenges and more like a fountain of fantastic opportunities. Of course, the task in front of us is monumental, but Ravi has a gift of following the predictable evolution of our past all the way into the future—placing it in the context of the increasingly internationally integrated environment that businesses will have to operate in to uncover rich opportunities that we can act on now.

Ravi strikes a unique and well-balanced tone in making his view of the world accessible to anyone. Big ideas are worthless without the right words to let the world know about them. Ravi's commitment to structuring his message into a format that is practical and actionable sets him apart from other visionaries, who often fall short when it comes to the "So, what now?"

Furthermore, Ravi and I have a shared appreciation for science, which we not only consider to be the tool to understand, dissect, and analyze the challenges at stake but also believe that, if this tool is properly mobilized, it will lead us to the very answers to these challenges. We know where to look for the solutions, and thus I share Ravi's profound conviction that we all need to urgently change the way we live and do business. Surely, the show must go on. However, it cannot and will not go on the same way. It's time for a new song. It's this new song that stands at the core of the training programs and conferences we have developed and launched together in recent years. It's also what *Twenty-First-Century Leadership to Fight the Code Red for Business* is all about.

Twenty-First-Century Leadership to Fight the Code Red for Business—and Ravi's work in general—is hopeful and empowering. At the consumer level, we've seen a rapid growth of a new type of mindfulness, a new sense of responsibility. Especially younger generations are adopting a new consciousness about sustainability, knowing that every dollar they spend or don't spend is a vote they cast for the world they want to live in.

Ravi's book takes this perspective and brings it to the office and the boardroom. Since the onset of this health crisis, even political and regulatory decision makers who were slow to act at first are now waking up to a new sense of responsibility. Like the individual consumer, leaders all over the world realize their power in turning up the pressure against unwanted externalities or in throwing their full support behind business models that can act as a force for good.

Twenty-First-Century Leadership to Fight the Code Red for Business shall serve them as both a source of inspiration and a trusted guide. The organizations that fail to understand this crucial shift and that are slow to align their models to these drastically changing market forces will not be around in five years. As much for businesses as for the planet, this is a matter of survival.

Surprisingly, far too many corporate actors still view sustainability as a loathsome burden, one linked to unreasonable regulations and costs. They remain blind to the huge opportunities that come with being a first mover, whose success Ravi lays out in compelling and practical terms. These remarkable winners, in B2B as much as in B2C businesses, stifle competition not only in serving their clients but also on the procurement side. They attract better financing *and* better talent. And they retain their talent. I am very excited to see that major universities and business schools are realizing that new business models also require a new kind of talent. Here again, the winners are those who have not only aligned their curriculum to new competencies but also adapted their teaching models to a new environment.

Yet as higher-education institutions have begun to step into the modern world, our primary and secondary schools are in many instances unable to follow. Wouldn't it be even more crucial in these formative years to prepare our students for the world that awaits them, the world they already live in? I know that Ravi's words will help to answer and lay foundations for this question too and open the doors for young people to become the twenty-first-century leaders that he elevates in his book. So, let's shape these leaders together and teach them to dance to a new song.

Raymond Schadeck
Chair Board of Universite dans la Nature (Canada)
Past President of ILA (Institute Luxembourgeois des Administrateurs) and Partner, EY Luxembourg

Twenty-First-Century Leadership to Fight the Code Red for Business makes a compelling case for twenty-first-century leadership. The accelerating technological, political, and social changes and the growing urgency to find a new balance with the natural environment pose new challenges for business leaders. Industrial-era mindsets are no longer sufficient to navigate the disruptions and uncertainties that define the new contexts within which businesses seek to survive and thrive.

Today's business leaders need to have the ability to recognize the vital signs of change and be informed by the best science available in the political, social, environmental, and technological domains. At the same time, they need to realize that the fundamental values of humanity remain unchanged. But interpreting change is not sufficient. Equally important is the ability to translate this understanding into strategy and operations.

Twenty-First-Century Leadership to Fight the Code Red for Business provides powerful examples of successful business leadership in the twenty-first century. It also identifies major technological, political, and environmental trends and thereby provides guideposts for the future. Hopefully, this work will inspire the next generation of business leaders. As the window of opportunity to avoid the code red for humanity is closing, the message of *Twenty-First-Century Leadership to Fight the Code Red for Business* is of utmost urgency.

Georg Kell
Founding Executive Director, United Nations Global Compact
Chairman, Arabesque Partners; Fund Manager, Anglo German ESG

Twenty-First Century Leadership to Fight the Code Red for Business by Dr. R. A. Fernando offers a fresh, much-needed, and exceedingly constructive perspective on the current and growing challenges facing our businesses, our nations, and our planet. His Twenty-First-Century Leadership Board Model takes a holistic view of our current crises—environmental, health and social, and economic (this last resulting from the first two)—and then, critically, considers the potential future disruptions, which will be caused by technology, geopolitics, or governance, for which leaders must develop both functioning radar and contingency plans.

Important and necessary as all this is, Fernando himself points out that many of these alarm bells have been rung repeatedly and urgently for years without sufficient impact. However, *Twenty-First-Century Leadership to Fight the Code Red for Business* and Fernando's vision provide three emphases that promise to move not only the conversation, but also the action, forward. First of all, he emphasizes leadership that is informed and guided by the most current insights from our best scientific minds but that also respects the tenets of the scientific method and the fact that as conditions change, our understanding and our behavior must change.

Too often in our quest for clarity, certainty, and predictability, we succumb to motivated reasoning and an all too easy dismissal of observations and insights that could throw our previous strategies into doubt. Such dismissals have led too many of us to underestimate the climate crisis and, more recently, to rush our return to something we think of as normalcy in the face of an ever-mutating pandemic. Both these responses—linked to an inadequate attention to both the science and the scientific method—have pushed our businesses, our communities, and our planet into great peril. Fernando's primary emphasis upon a leadership that respects science is essential for our survival.

The second emphasis that distinguishes Fernando's book and takes it beyond another valuable but too often ignored warning bell is his unfailingly constructive, practical, action-oriented approach. It is this part of Fernando's mindset and personality that interested me when we first met years ago at INSEAD. He not only talked about values-driven leadership in the abstract but also had many stories of his own successful efforts to enact this sort of leadership drawn from his corporate experiences—and his stories were not self-aggrandizing tales but, rather, detailed explanations of the strategies, tactics, interpersonal insights, scripts and arguments, and team lessons that enabled positive impacts.

This practical and positive action orientation is a key component of Fernando's book. Throughout the text, he includes corporate case examples to illustrate not only the importance and feasibility of a science-based, sustainability-focused approach to organizational leadership but also the many paths one may take and the many tactics one may use to achieve it.

Finally, the third emphasis that distinguishes Fernando's work is the importance placed on developing a corporate sustainability *strategy*. That is, while acknowledging the impossibility of a totally predictable future, he argues for the importance of attending to a variety of potential future scenarios informed by the current observable seeds of disruption—whether they are technological, geopolitical, or governance based—and then building a sort of *contingency resilience* to mitigate risks that may emerge.

That is, rather than succumbing to the lure of assumed certainty and predictability—a certainty that too often masquerades as confidence and strength—Fernando argues for leaders who consider the best information at hand but, as true scientists would, also continue to view that information as hypotheses that must be tested and retested. And what's more, he argues for the type of leadership that can embrace humility—knowing that one doesn't know what one doesn't know—while simultaneously building organizations and strategies that are nimble, able, and willing enough to alter course when conditions necessitate.

In a world that is increasingly at risk from environmental and ideological threats, Fernando's call and blueprint for committed, informed, and strategic leaders—accompanied by concrete examples of successful tactics—is desperately needed.

Mary C. Gentile, PhD
Creator/Director, Giving Voice to Values
Formerly the Richard M. Waitzer Bicentennial Professor of Ethics, University of Virginia Darden School of Business
www.GivingVoiceToValues.org and www.MaryGentile.com

As a professional dedicated to sustainability and being tasked with leading a national center of excellence for low carbon in Ireland, I consistently endeavor to understand and draw inspiration from sustainability best practices and opinion from thought leaders across the globe.

It was in this context that I first met Dr. R. A. Fernando. While navigating and interacting with mutual connections within our professional networks, we somewhat serendipitously were prompted to connect. Casual at first, Ravi sitting in a tea shop in Sri Lanka and I in a coffee shop in Dublin, we began to discuss the climate issues of our time. From the humble beginnings of a Skype call sipping hot drinks—the similarities in our surroundings somewhat symbolic of the relatedness in our opinions—came an enduring relationship and, later, a tangible application of Ravi's extensive knowledge and experience being implemented across the Irish business community through his training, of which this literature is a core pillar.

Twenty-First-Century Leadership to Fight the Code Red for Business offers the reader insight into some of the most contentious and troubling concepts about climate change and the core instigators and enablers of our planet's increasing temperature. Riveting, upsetting, but always informative, *Twenty-First-Century Leadership to Fight the Code Red for Business* will tell you the things mainstream media and governments won't.

Niall Kelly
Managing Director, the Cube, Ireland

Humanity came together at the end of the twentieth century to explore its future. The United Nations report *Our Common Future* spelled out in 1987 for the first time the biggest challenges we are all facing. It clearly stated that we are compromising our future generations' ability to meet their own needs and that the global pace of use of resources is too high to sustain and poses massive challenges for the future.

Humanity acted on the report. In 1992 at the Rio Conference, world leaders declared intentions to become sustainable. Statements were made and plans outlined. However, more than three decades later, the successes are very limited. While we have grown our population by more than three billion since then and, having managed to address malnourishment and hunger relatively well, reduce absolute poverty, and achieve some meaningful progress in terms of basic education and infant mortality, we have largely failed to act decisively and effectively on our sustainability challenges. We have compromised and are continuing to compromise future generations' ability to meet their own needs. This is most evident in the area of climate change.

The world has warmed by more than one degree Celsius compared to preindustrial times and is clearly on a trajectory to exceed the limits of what is considered to be manageable as globally agreed in the goals of the Paris Agreement. We are already suffering the consequences and will increasingly continue to do so.

The responsibility to ensure the sustainability of our future is undoubtedly shared among governments, businesses, and citizens. However, businesses have a particularly critical role to play. Their ability to develop products and services that cater purely to the shareholders and their profits but to do so while taking into account the wider needs of society and the environment is paramount. It is well understood that a company cannot survive if it does not make a profit, but it cannot survive in the long term if it only makes a profit. Its competitiveness—more fundamentally, its license to operate—is dependent on society and, with it, its ability to cater to the people and the planet. No company can succeed in a failed society. Integrating sustainability into the business strategy and placing it at its heart is key to a company's competitiveness. Every company wants to achieve success in the short term, but also in the long term. Long-term success starts in the present and incorporates the needs of tomorrow.

The twenty-first century needs a new model of good leadership. This seminal book on twenty-first-century leadership clearly spells out the challenges and emergencies ahead and describes the key dimensions that every board member has to take into account to be a successful business leader, navigating the company toward long-term success. It provides a model needed for board members going forward, encompassing an approach where the emergencies that need to be tackled are outlined and solutions are provided.

Business as usual will not do. Holistic and science-based thinking with a long-term view, connecting the business' activities with their material effects on people and the environment, as well as the environment's effects on the company, must be part of the backbone of the twenty-first-century strategy. There is no more time to wait.

Dr. Hakan Lucius
Head of Corporate Sustainability,
European Investment Bank, Luxembourg

Dr. R. A. Fernando is as relentlessly uncompromising as he is pragmatic where the climate emergency is concerned. His call for immediate action and science-based adaptation is as important for the commercial survival of the business as it is for the existence of communities threatened by the growing health, humanitarian, and environmental challenges that already confront us. This is a guide for the enlightened businesses that recognize the moral, ethical, and commercial imperative to realign for the future in delivering a habitable earth for our sons and daughters.

Dilhan C. Fernando
CEO of Dilmah Ceylon Tea, MJF Charitable Foundation, and the Dilmah Conservation

Dr. R. A. Fernando has always shown great dedication to the climate emergency. He demonstrates a refreshing outlook, combining a vast knowledge of the international business world and determination to find sustainable solutions. We are excited to see his revolutionary business model in action, a beacon of hope for our leaders of the future as they confront the climate crisis.

Carl Craen
Vice President & Managing Director, EU Business School

ACKNOWLEDGMENTS

A sincere word of appreciation to the following individuals for contributing a foreword:

Georg Kell, chairman, Arabesque Partners, and founding executive director of the UN Global Compact

Raymond Schadeck past president of ILA (Institut Luxembourgeois des Administrateurs), and chair, Universite dans la Nature

Professor Mary Gentile, creator/director of Giving Voice to Values; formerly the Richard M. Waitzer Bicentennial Professor of Ethics, University of Virginia Darden School of Business

Dr. Hakan Lucius, Head of Corporate Sustainability, European Investment Bank, Luxembourg

Niall Kelly, managing director, the Cube, Ireland

Dilhan C. Fernando, CEO, Dilmah Ceylon Tea, MJF Charitable Foundation, the Dilmah Conservation

Carl Craen, Vice President & Managing Director, EU Business School

Amjad Azmeer, director of research at Global Strategic Corporate Sustainability Pvt. Ltd., for the extensive research and mini case summaries that have significantly enhanced the credibility and veracity of *Twenty-First-Century Leadership to Fight the Code Red for Business*

Navindi Wellmillege of Design Nariya, for the cover design and illustrations for *Twenty-First-Century Leadership to Fight the Code Red for Business*

TWENTY-FIRST-CENTURY LEADERSHIP TO FIGHT THE CODE RED FOR BUSINESS

In 1987, the Bruntland report titled *Our Common Future* was presented to the United Nations. The impending climate emergency was brought to focus with the urgent need to end reliance on pollution-causing fossil fuels (coal, petroleum, and natural gas) as the key solution.

Between 1987 and 2022, for thirty-five years, world, national, and business leaders have ignored science and the urgent need for climate action such as transitioning away from fossil fuels. Despite many UN-led global forums, climate action failure and extreme weather were identified by the World Economic Forum in 2022 as the two key risks facing the planet as a result of the failure of science-denying leaders to address the global risks of having continued to rely on fossil fuels for the past thirty-five years. Such reliance on fossil fuels has resulted in the code red for humanity and business.

The UN Earth Summit in Rio, Brazil, in 1992, the UN Kyoto Protocol in Japan in 1997, the UN Paris Agreement in 2015, and the UN's COP26 (Congregation of Parties 26) meetings have all failed to compel leaders to take focused action to end their dependence on fossil fuels. At every one of these meetings, the date to finally phase out fossil fuels was postponed, the can kicked down the road.

In 1992 at the Rio Earth Summit, the world talked of ceasing its reliance on fossil fuels by 2000. The date was shamelessly postponed again, this time until 2030, at the meeting in 2015 to sign the Paris Agreement. At the COP26 in 2021 in Glasgow, science-denying leaders postponed taking action yet again, stretching the date to between 2050 and 2070! Does this show an urgency to respond to the climate emergency? This is catastrophic climate action failure.

The exception to the rule is a few enlightened, science-led, twenty-first-century leaders with a sustainability mindset. The leaders of Denmark, Bhutan, Costa Rica, and South Australia, and business leaders Elon Musk (Tesla), Sundar Pichai (Google), Yvon Chouinard (Patagonia), Werner Hoyer (EIB), and Henrik Poulsen (Ørsted), to name a few, are these exceptions to the rule. Their numbers are too few to halt the catastrophic destruction of the planet. Today, we burn significantly more fossil fuels now than we did in 1987.

Nations exporting fossil fuels; oil and gas companies; and banks that funding fossil fuel corporations are led by nineteenth-century science deniers who all invest in fossil fuels, using their financial strength and influence to lobby for the delay of any decisive action to transition away from fossil fuels. Guilty parties include Saudi Arabia, Russia, China, India, Turkey, Australia, Mexico, Brazil, the United States, and Canada. In fact, we have burned more fossil fuels since 1987, resulting in an increase of carbon dioxide emissions from 20 gigatons to 22 gigatons in the 1980s to a concentration of more than 33 gigatons of CO_2 in the present day—up from 280 ppm to 420 ppm in that period—putting the planet on a path toward extreme-weather-led devastation and a code red for humanity and business.

The book *21 Lessons for the 21st Century* by Yuval Noah Harari inspired me to consider some of the key lessons that I should present. This is in addition to being a sustainability-mindset leader who demonstrates the four traits of twenty-first-century leadership in order to be relevant and strategic. The Twenty-First-Century Board Leadership Model is a combination of all these ideas and concepts.

The conceptualization and creation of the Twenty-First-Century Board Leadership Model© in June 2020 focused attention on the three **emergencies** in the first half of the model, which covers the climate emergency, the health and social emergency, and the resultant economic emergency.

The second half of the model focuses on the three **disruptors**—technology, geopolitics, and governance—that need strategic action. This led to the creation of the Twenty-First-Century Board Leadership Model MasterClass© in September 2020, which was launched in Luxembourg by the Institute Luxembourgeois des Administrateurs.

The four key traits that set apart a twenty-first-century leader are as follows:

- an appreciation for mobilizing science as part of the business strategy
- an urgent need to make an impact on the code red for business
- a sustainability mindset that prioritizes the planet and people over profit
- a willingness to embed the Twenty-First-Century Board Leadership Model in the business's board/C-suite/operations agenda.

We can either ignore science, as the nineteenth-century-mindset leaders have done by placing profit before the planet for the past thirty-five years, or become relevant, twenty-first-century, science-led, sustainability-mindset leaders who feel an urgent need to make an impact on the code red for humanity and business. We need to be a key part of the solution instead of simply continuing with business as usual. We need to do all we can by 2025, and not a day later, to save the planet.

The goal of *Twenty-First-Century Leadership to Fight the Code Red for Business* is to create a cadre of twenty-first-century leaders ready to fight the code red for business by, first, ceasing to use fossil fuels for energy, electricity, and transport and thereby reduce the organization's carbon footprint in terms of Scope 1, 2, and 3 emissions to be part of the solution to the climate emergency.

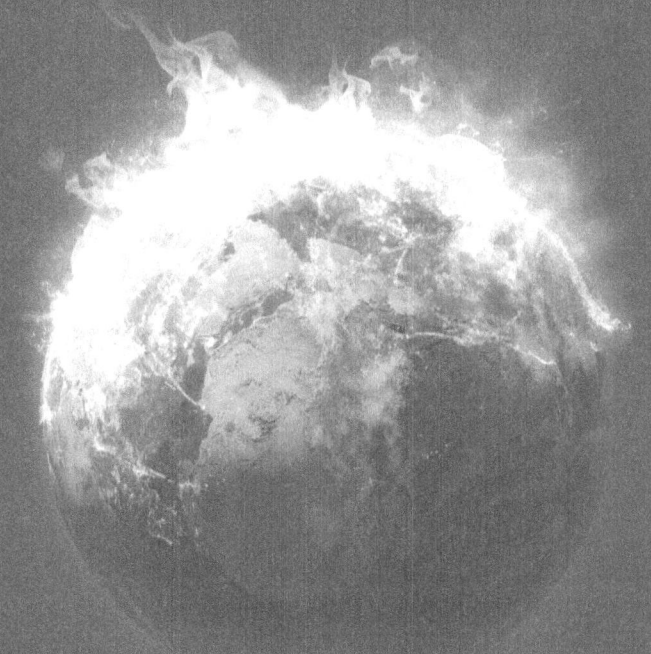

CHAPTER 1

IS THERE A CODE RED FOR HUMANITY AND BUSINESS?

CHAPTER 1

IS THERE A CODE RED FOR HUMANITY AND BUSINESS?

Today we live in a world where the words *truth*, *fact*, *science*, *reality*, and *emergency* are all compromised and controlled by interest groups. We have a world where leaders of nations, businesses, and media are controlled by a few global institutional investors whose boards are populated by billionaires with significant investments in global industries that pollute the planet, such as the fossil fuel and plastic industries. These individuals would be the first to object to ending our reliance on fossil fuels and have used their global influence to present a version of the truth acceptable to the interest groups. Today we live in a world where the media, controlled and influenced by the very same organizations, have a monopoly on the version of the truth that meets the objectives of a few, versus disseminating the truth in the real sense of the word.

In the case of the climate emergency, at COP26 in Glasgow, Scotland, we saw leaders of Saudi Arabia, Russia, China, Australia, India, Turkey, Mexico, and Indonesia present a version of truth that served the agenda of profit over the planet in terms of transitioning away from the use of fossil fuels.

In terms of the health emergency, the USA, Brazil, India, and Russia have all placed profit over people, and to date these countries remain the top nations for COVID-19 deaths. It must be noted that the USA rejoined the Paris Agreement as guided by the advice of its health adviser starting in January 2020 during the Biden administration.

The impact of the beliefs of a few science-denying, truth-denying leaders exemplifies what is happening in the world today. The words *facts* and *science* have been ignored in the world for the past thirty-five years. Those invested in fossil fuels have presented a version of the truth that has kept their investments thriving, delaying any climate action. The meaning of the word *emergency*—often used in the ICU (intensive care unit) and ERU (emergency response unit) to signal when something requires immediate and comprehensive attention—has completely lost its meaning. This has been replaced by a version of the truth that now makes the words irrelevant as we fight the code red for humanity and business.

Nineteenth-century, science-denying, short-term-focused leaders invested in fossil fuels have deflected the need to address the climate emergency. They have procrastinated and postponed taking decisive action since 1987 by pushing back the date to transition away from fossil fuels at every given opportunity. In 1992, when the UN Earth Summit in Rio took place, the world wanted to end its reliance on fossil fuels by 2000. In 2015 at the talks for the UN Paris Agreement, this date was postponed to 2030. At COP26, this date was postponed irresponsibly to 2050 or 2070. The only factor considered is how the required actions impact the status quo and disrupt business as usual. Transitioning away from fossil fuels can be put off by doing as little as possible and making frivolous commitments to do so by 2050 or 2070, mentioning such strategies as deploying untested technologies such as carbon capture and carbon offsetting.

While the planet will face an existential threat in the period of 2025–30, as confirmed by both the World Meteorological Organization (WMO) and Intergovernmental Panel on Climate Change (IPCC) reports published in 2021–22, the decision to take on climate change by ceasing to use fossil fuels has been postponed irresponsibly. The next time you hear a version of the truth from the media, do not believe it. Instead, check it for the scientific veracity and reality.

On May 17, 2021, BBC News confirmed that the World Meteorological Organization reported that we could see the planet reach a tipping point by warming up from +1.5°C to +2°C between 2025[1] and 2040. This is from the base year 1880, when the Industrial Revolution began, as did the burning of fossil fuels—coal, oil (petroleum and diesel), and natural gas. The IPCC reports from 2018 to 2021 (especially AR6, the Sixth Assessment Report) confirm that the global economic fallout from rising sea levels could cost between US$54 trillion and US$67 trillion in a $100 trillion global economy as of 2022. We are also staring at a new tipping point of a temperature change to +3.7°C, which could result in economic devastation to the tune of US$515 trillion on account of a 140-foot sea level rise by 2040.

The effect of crossing planetary boundaries is discussed by both Professor Johan Rockström and Sir David Attenborough in the Netflix documentary *Breaking Boundaries*,[2] which mentions that we have already crossed four to five of the nine planetary boundaries. This was an alarm bell that most science-denying leaders did not respond to with the urgency it deserves. The Netflix movie *Don't Look Up*[3] confirms beyond a reasonable doubt that most science-denying world leaders and science-denying business leaders have not taken the climate emergency and code red for business seriously. In fact, they have ignored it despite all the warnings. The film also raises the question of how many of these leaders are science led. The answer presented by most nineteenth-century science-denying leaders to the question of catastrophic climate events seems to be "Don't look up"—in other words, simply ignore the scientific evidence and reality. This is now the emerging trend and phenomenon.

Code Red for Humanity and Business—Our House Is on Fire

Greta Thunberg, the young climate activist, said it clearly at COP25 in 2020:[4]

> We no longer have time to leave out the science. The global carbon budget available before we hit catastrophic

global warming of +1.5°C from the 1880 base year. The rapidly declining carbon budgets are referred to in the IPCC (Intergovernmental Panel on Climate Change) report of 2019, if we are to have a 67 percent chance of limiting the global temperature rise to below 1.5°C. On January 1, 2018, we had a carbon/GHG [greenhouse gases] budget of 420 gigatons. At the global average emission of 42 gigatons of CO_2 per year including land use. As at 2020, this budget declined to 340 gigatons CO_2/GHG. At the 42 gigatons of consumption we will exhaust the entire budget in eight years from *now*! The implication is that from January 2022 we have till January 2030 at best as with each passing year we emit more due to excessive burning of fossil fuels.

As per the current best available science, this requires the major coal-reliant, fossil-fuel-exporting nations and high-GHG-emitting nations—China, India, the USA, Japan, Turkey, Russia, Saudi Arabia, Mexico, Australia, South Korea, Poland, and Germany—to transition away from using pollution-causing fossil fuels by 2025 in order to give the planet's lesser-developed nations a chance to do the same by 2030. Now we are staring at a global warming scenario of +1.5°C to 2.7°C/3.7°C, which will result in global economic devastation. As per the IPCC report, if we cross the +3.7°C level, we will see a 140-foot sea level rise, which would lead to the devastation of the US$515 trillion of global GDP, wiping out nearly five years of cumulative GDP.

Despite our house being on fire, the fossil fuel industry continues to burn more fossil fuels, and irresponsible banks have invested $1.9 trillion to $3 trillion in fossil fuel projects since the 2015 December Paris Agreement, where they committed to halt such investments. G20 nations account for 80 percent of the emissions; one hundred companies are responsible for 71 percent of emissions; and the richest 10 percent account for 50 percent emissions—while the poorest 50 percent account for just 10 percent.

Where is the urgency to transition away from using fossil fuels? At COP26 (a cop-out for the twenty-sixth time), we saw the nations most responsible for 92 percent of cumulative emissions of CO_2 to date making irresponsible commitments to transition away from fossil fuels between 2050 and 2070. Where is the urgency to address the climate emergency and the code red for humanity and business?

Sea Level Rise and Economic Devastation[5]

The prognosis is that we will cross the climate change barrier by 2025 to 2040 and reach between +2.2°C and +2.7°C by 2030, which will make a catastrophic impact on the globe and give rise to extreme weather incidents, extreme temperatures (40°C–55°C), droughts, floods, typhoons, and the unprecedented acceleration of the melting of the polar ice caps.

The BBC program titled *Life at 50°C*[6] is a warning of the climate emergency the planet is hurtling toward. The indecision and failure of our leaders in the past thirty-five years could make this a reality sooner rather than later.

As excessive burning of fossil fuels increases and GHG/CO_2 concentrations increase, we are seeing the accelerated melting of the Arctic, Antarctic, and Himalayan ice caps because of global warming. As per IPCC and WMO, the forecasted twenty-foot sea level rise is projected to devastate the global economy.[7]

The reality is that science-denying "leaders" have not responded to this scenario.

Can we expect these leaders to respond to these threats?

According to Burke, Hsiang, and Miguel (2015),[8] we will see the following impacts as a result of sea levels rising:

- A majority of nations in the Southern Hemisphere are projected to lose between 50 percent and 100 percent of GDP by 2100, especially Mexico, Latin America, Africa, the Middle East, India, and Southeast Asia.

- North America, Russia, and European nations are going to be less affected by the climate emergency than Southern Hemisphere nations.

The World Economic Forum (WEF) in Davos has corroborated the fact by identifying the top five global risks most likely to happen in both 2020 and 2022 and listing extreme weather and climate action failure as the top two global risks in all three years.

Following is a list of global risks we will face as identified by the Global Risk Report 2022 and the World Economic Forum:[9]

1. climate action failure
2. extreme weather
3. biodiversity loss
4. social cohesion erosion
5. livelihood crises
6. infectious diseases
7. human environmental damage
8. natural resource crises
9. debt crises
10. geoeconomic confrontation.

What is clear is that the world leaders at the World Economic Forum (WEF) sessions in 2020, 2021, and 2022 all agreed that extreme weather and climate action failure are the two key global risks set to devastate the global economy, this despite an obsession to focus attention on growing the global economy at any cost. What does the 2020–22 WEF ranking of the two top global risks imply?

1) **Climate action failure**
   This confirms the hypothesis that since the 1987 Bruntland report, world leaders, business leaders, and all those in positions of influence have had a nineteenth-century, science-denying mindset and have ignored the scientific evidence and the urgent

need to cease using pollution-causing fossil fuels for thirty-five years (1987 to 2022).

Despite many highly visible and publicized events orchestrated by the United Nations between 1992 and 2022, including twenty-six COP (Congregation of Parties) meetings between 1996 and 2021, zero action has been taken to solve the problem that has caused the code red for humanity and business and the climate emergency.

2) **Extreme weather**

The end result of the catastrophic leadership and climate action failure over the past thirty-five years, including the failure to transition away from pollution-causing fossil fuels, has been extreme weather events that have devastated global economies and created a new cadre of climate/environmental refugees at an unprecedented rate. From fewer than ten extreme weather events in the 1970s to more than four hundred incidents per annum by 2022, this has resulted in a loss of seventeen billion US dollars per annum in the 1970s to more than five hundred billion US dollars per annum in 2022, a figure that is on a course to rise even further as we cross the +1.5°C to +2.4°C global warming tipping points between 2025 and 2030.

In addition to the aforementioned two global risks, **biodiversity loss** (once again the result of a leadership failure to halt deforestation and reforest the planet) has laid the foundation for pandemics and the current health emergency. Nations like Brazil exemplify this fact. Brazil's Amazon rain forest, the earth's lung, used to account for 20 percent of the global oxygen supply, but today, after President Jair Bolsonaro's devastation of the rain forest, we see it is a zero-oxygen contributor and now a key part of the problem. This came about as a result of biodiversity loss, and the indirect impacts on the next three social sustainability global risks—social cohesion erosion, infectious diseases, and livelihood loss—will be exacerbated.

Human environmental damage continues with devastating oil spills, deforestation in the name of agriculture, and forest fires set off by a warming planet. COVID-19, which has infected four hundred million people and killed more than six million, has wakened the world to the reality of infectious diseases and their devastating effects. Finally is the loss of biodiversity, which has given rise to global animal-led pandemics (e.g., SARS, Ebola, H1N1, MERS, and COVID-19). By 2021 we had destroyed more than 75 percent of our global rain forests and, with them, more than 60 percent of earth's species. On closer examination, these facts point us in the direction of catastrophic leadership and climate action failure of unprecedented proportions. At the UN General Assembly sessions in 2021 and at COP26 in Glasgow, Scotland, the UN secretary-general António Guterres called the situation a "code red for humanity." A few years earlier, young Greta Thunberg of Sweden called on world leaders, saying, "Panic! Our house is on fire."

The Code Red for Humanity Is a Code Red for Business

Despite all the warnings since 1987, we see that a fossil fuel lobby made up of the nations that are reluctant to address the climate emergency and code red for humanity because of their reliance on coal and petroleum has influenced the postponement of decisive action to cease using the very things that have caused the climate emergency—fossil fuels (coal, petroleum, and natural gas) to between 2050. Nineteenth-century science-denying nonleaders in business are complicit in helping to cause the code red for business and humanity. As reported by CNN in September 2021, leading up to COP26, the following nations had made no commitments to reduce emissions by 2030 prior to COP26.

- Australia
- Brazil
- China
- India
- Indonesia

- Mexico
- Russia
- Saudi Arabia
- Turkey.

At COP26, most nations made commitments to transition away from fossil fuels at a time long after we are projected to cross the +1.5°C to +2°C tipping point, between 2025 and 2040. In fact, the majority of the commitments made were for between 2050 and 2070. We also know that Saudi Arabia, India, Australia, Russia, and India objected to any mention of transitioning away from fossil fuels in any of the resolutions, agreements, policy documents, or declarations because they are the nations that rely on fossil fuel exports. This confirms that these nations, along with fossil fuel companies, are prioritizing profit over the planet. If our house is on fire, will we call 911 sometime between twenty and fifty years after the fact? The failure to address climate change is science denial at its worst! It is happening despite the fact that the planet has recorded its hottest ever twenty-two years since 2000.[10]

Are Extreme Weather Incidents Increasing as a Result of Global Warming?

Following is an analysis of the frequency of extreme weather incidents, their impact on the global economy, and deaths caused by the climate emergency. World leaders have turned a blind eye to the increase in extreme weather incidents, from an average of seven per annum in the 1970s to more than three hundred fifteen per annum by 2019—and an off-the-charts four hundred to five hundred extreme weather incidents in 2021.

The accelerating frequency of extreme weather events as per visual capitalist data streams confirms that droughts will increase 4.1-fold; extreme rainstorms will increase 2.7-fold; and heat waves will increase 9.4-fold when we cross the +4°C mark, confirming the impact of the burning of fossil fuels.

The burning of fossil fuels has led to the devastation of the global economy, from US $18 billion in the 1970s to more than US $138 billion per annum by the 2000s. In 2020–21, we saw this figure rise to between US$250 billion and US$300 billion per annum. The number of lives lost on account of extreme weather incidents and the climate emergency is estimated to be more than two million to three million in the period from 1970 to 2021. This does not take into account the massive loss of seven million lives—people who died because of air pollution caused by the burning of fossil fuels. Therefore, we could say that between ten million and twelve million people have died because of the climate and health emergency.

The World Meteorological Organization's *Atlas of Global Economic Losses and Mortality from Weather—Climate and Water Extremes (1970–2019)*[11] catalogues the major impacts and the economic devastation with authority and credibility.

1. Extreme weather incidents
2. Economic losses
3. Loss of life

The number of extreme weather incidents per annum increased from seven hundred eleven in the period 1970–79 to more than three thousand per annum between 2000 and 2019. If we look at 2020 to 2022, we see a further increase in both intensity and frequency in the more than four thousand extreme weather incidents as the planet recorded its hottest seven years in history.

The devastation that these extreme weather incidents have on the global economy has also been dramatic, from US$175.4 billion between 1970 and 1979, to a tenfold increase between 2000 and 2019, to between US$942 billion and US$1.381 trillion from 2010 to 2019. What is most disturbing is that the intensity of these weather incidents has increased too. For 2020–22, the figure moved up significantly, creating havoc in the global economy.

Finally, we look at the loss of life. In the period 1970 to 2019, we see that more than two million human beings lost their lives. One needs to add the seven million deaths per annum owing to air pollution as per the WHO reports.

The facts are available for all to see, and they have been available since 1987. Given that we know the cause, why haven't we ceased using pollution- and global-warming-causing fossil fuels since 1987? How long have world leaders known that the burning of fossil fuels is what has led to the climate emergency? Has the word *emergency*, used by the medical profession to ensure lives are not lost by urgent and focused medical interventions in the emergency trauma centers, intensive care units, and cardio care units, lost its meaning because leaders have been ignoring the climate emergency and the code red for humanity?

Have world leaders consistently prioritized profits over the planet, ignoring the need to transition away from using pollution-causing fossil fuels, having been influenced by the corrupt fossil fuel and banking industries? Have world leaders and business leaders compromised humankind today and future generations by their flippant disregard for science and for the urgent need to fight the climate emergency? Has the phrase *code red for humanity*, as a result of the climate emergency, been ignored and action required postponed from 1987 to 2021, or 2040, or 2050?

If in any business in which you were employed you were to miss your targets for thirty-five years in a row and, not only that, actually contribute to a catastrophe where the issue that you were supposed to address had worsened by more than 50 percent, would you still have a job? Would your organization survive?

Chapter 1—Twenty-First-Century Implications

1) The 1987 Bruntland report confirms that the climate emergency was indeed human induced and that we urgently need to end our reliance on fossil fuels.
2) Between 1987 and 2022, a majority of world leaders and business leaders have ignored the science and thereby failed to take decisive action to cease using pollution-causing fossil fuels—an instant of climate action failure with catastrophic consequences.

3) The end result of the excessive burning of fossil fuels for the past thirty-five years is global warming. The past twenty-two years since 2000 have been the hottest on record.

4) Extreme weather incidents are precipitated by global warming. From a base year of 1880, we have reached +1.2°C, and we expect the frequency of extreme weather incidents and the economic devastation that accompanies them to increase as we cross the +1.5°C and +3.7°C tipping points between 2025 and 2040.

5) We have seen a devastating increase in extreme weather incidents, from fewer than ten per annum between 1970 and 1979 to more than three hundred incidents between 2000 and 2019. In the past few years, 2019 to 2022, this figure rose to more than five hundred incidents per annum. Extreme temperatures, droughts, typhoons, hurricanes, and floods are part of the new normal. Flooding accounts for more than 70 percent of all extreme weather incidents.

6) The economic losses have increased from twelve billion US dollars per annum in the 1970s to more than one hundred fifty billion US dollars per annum between 2000 and 2019. In the past few years, 2019 to 2022, we have seen this figure rise to more than four hundred billion US dollars per annum.

7) The UN secretary-general has referred to the climate crisis as a "code red for humanity."

8) Extreme global warming is melting the Arctic, Antarctic, and Himalayan ice caps, leading to an unprecedented rise in sea level and economic devastation for most coastal cities in the world.

9) The rise in sea level is forecasted by the IPCC and WMO to be between twenty feet and one hundred forty feet as we cross the +1.5°C to +3.7°C tipping point between 2025 and 2040.

10) The estimated economic devastation that world leaders ignore is estimated by the IPCC to be between US$54 billion and US$67 trillion in a US$100 trillion global economy in 2022. As per the World Bank's Groundswell report, there will be 143 million climate refugees by 2050.

Melting
of Polar Caps

CHAPTER 2

THIRTY-FIVE YEARS OF SCIENCE DENIAL—A FAILURE OF CLIMATE INACTION FROM 1987 TO 2022

CHAPTER 2

THIRTY-FIVE YEARS OF SCIENCE DENIAL— A FAILURE OF CLIMATE INACTION FROM 1987 TO 2022

If in 1987 world leaders knew the impending climate emergency was as a result of the burning of pollution- and global-warming-causing fossil fuels, why did they fail to act with resolve and urgency to transition away from fossil fuels? Instead, they increased the gigatons of atmospheric CO_2 from 21 to 33.9 and have subsidized fossil fuels for the past thirty-five years. This is climate action failure.[1]

In 1992 the UN launched its Earth Summit in Rio, Brazil, after a lull of five years since the Bruntland report. Apparently, one hundred ninety or so world leaders and business leaders were convinced that the time for action was now and made "talk is cheap" commitments to reduce their dependency on fossil fuels by 2000.

In 1997 the Kyoto Protocol once again highlighted the need to end our dependence on pollution-causing fossil fuels and cut down GHGs (greenhouse gases), specifically, carbon emissions, by both developed and developing nations making deep cuts in their usage. Instead of moving decisively on the matter, those at the meeting argued about who should be the first to make such reductions.

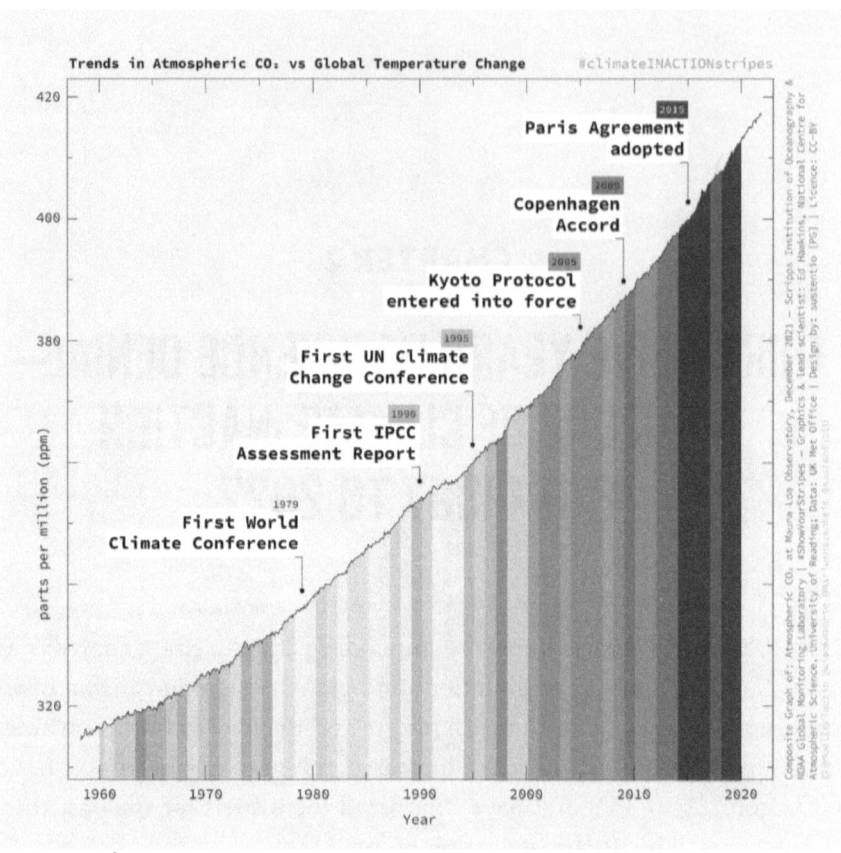

Climate Emergency—Thirty-Five Years of Science Denial and Climate Action Failure[2]

In 1987, the World Commission on Environment and Development published *Our Common Future*,[3] also known as the Bruntland report, which left absolutely no doubt the planet was in a state of emergency in terms of the climate because of the continuing use of fossil fuels. In June 1988, James Hansen[4] of the NASA Goddard Institute presented his testimony to the US Congress confirming that the unfolding climate emergency was man-made and needed immediate action in the form of ending our dependence on pollution-causing fossil fuels.

The UN-led COP (Congregation of Parties) had its first meeting in 1996 and its twenty-sixth meeting in November 2021 in Glasgow, Scotland. None of these meetings has ever ended with a decision to cease using fossil fuels with urgency. Since 1987, though many global forums and meetings have been held, some of which follow, climate inaction is all we have seen:[5]

- the 1992 Earth Summit in Rio, Brazil
- the 1997 Kyoto Protocol in Japan
- the 2015 December Paris Agreement in France
- twenty-six COP meetings in various global cities from 1996 to November 2021.

Limiting global warming to *below a 1.5°C increase* from the preindustrial base year of 1880 has been the UN's goal. The way to achieve this is by ending our reliance on pollution-causing fossil fuels. How successful has the UN been for the past thirty-five years?

Nineteenth-century, science-denying, code-red-for-humanity-denying, climate-emergency-denying leaders have extended the time line to end dependency on pollution-causing fossil fuels. Thirty-five years of denial and climate inaction have led to a postponement of action until 2050, 2060, or 2070. In reality, the indecision and failure to act for thirty-five years and at COP26 has accelerated the climate emergency, as demonstrated below, despite the many UN meetings between 1987 and 2022.

Despite all the aforementioned global meetings, summits, conferences, accords, and agreements, the 1987 Bruntland report has not resulted in decisive action to end our reliance on pollution-causing fossil fuels, which have caused global warming, the climate emergency, and the code red for humanity and business. The end results are as follows:

1) We have burned more fossil fuels.
2) We have not ended our reliance on fossil fuels.
3) We subsidize fossil fuels.
4) The planet has grown warmer by +1.2°C with each passing year.
5) The amount of the greenhouse gas carbon dioxide in our atmosphere has now reached 33.9 billion gigatons and 420 ppm.

The foregoing facts beg the question of whether the global, national, and business leaders who attended these meetings for the past thirty-five years nineteenth-century science-denying leaders who ignored the most important action required, namely, ending our dependence on fossil fuels? The fossil fuel companies and banks led by nineteenth-century science-denying nonleaders thrive while the planet faces a code red for humanity and business!

Which nations have contributed most to the climate emergency? The Global North is responsible for 92 percent of the problem. The onus is clearly on the Global North—USA, Europe, Russia, and Australia—to be the first to act with urgency to end our reliance on pollution-causing fossil fuels and fund the energy transition of the Global South (China, India, and the rest of the world), which is only 8 percent responsible for the climate emergency.

At COP26, we saw a cop-out of great proportions considering the fact that more than 92 percent of global year-to-date emissions[6] are from the Global North, the nations clearly called out for their lack of urgency to respond the climate emergency. These include the major fossil-fuel-exporting or -using nations—Saudi Arabia, China, Russia, Australia, India, Mexico, Indonesia, and Brazil. What is also noteworthy is that these nations are also the ones rated as "critically insufficient" or "highly insufficient" in their readiness to fight the code red for humanity.

In 2000, Kofi Annan, UN secretary-general at the time, launched the Millennium Development Goals[7] (MDGs) in order to galvanize the UN signatory nations to impact the sustainability of the planet from 2000 to 2015. Just three MDGs were achieved, thanks to China.

Then came the December 2015 Paris Agreement,[8] where 196 nations committed to transition away from pollution-causing fossil fuels by 2030. Each nation made a commitment, referred to as nationally determined contribution (NDC), to move to 80 percent renewable energy by 2030 and achieve the seventeen UN Sustainable Development Goals. A review of the progress in terms of the NDCs in the UNEP Gap Emissions Report showed most nations were more than 70 percent off the target they should have delivered by 2020.

To achieve the 2030 commitments, world leaders need to first achieve the missed 70 percent for the period 2015 to 2020,[9] and thereafter achieve 100 percent of what needs to be achieved between 2023 and 2030—the next seven years! It is increasingly clear that this is impossible, which once again raises the question, were the leaders who made these commitments nineteenth-century science deniers?

If one looks at the three key parameters that matter, one ought also look at how world leaders have failed on all three counts:[10]

1) gigatons of CO_2/GHG—1987, 21; 2021, 33.9;
2) carbon concentrations from in the preindustrial period under 280 ppm to 420 ppm in 2021;
3) an increase in temperatures from 1880, the preindustrial period, by 1.2°C.

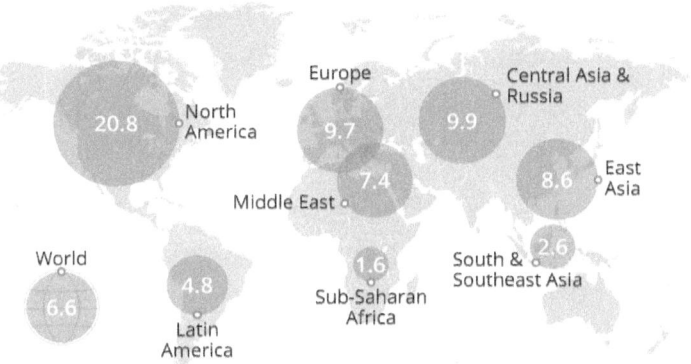

Humanity's Uneven CO₂ Footprint

Average CO_2 emissions per capita in selected regions in 2019 (in tons)*

20.8 North America
Europe 9.7
Central Asia & Russia 9.9
7.4 Middle East
8.6 East Asia
World
6.6
4.8 Latin America
1.6 Sub-Saharan Africa
South & Southeast Asia 2.6

* Including emissions of domestic consumption and net imports of goods & services
Source: World Inequality Report 2022

statista

Humanity's uneven CO_2 footprint[11]

FIGURE 1
2019 net GHG emissions from the world's largest emitters
Million metric tons of CO₂e, including emissions and removals from land-use and forests and share of global total

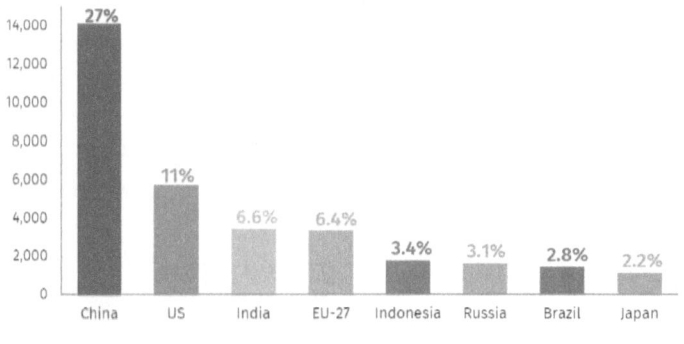

Source: Rhodium Group

2019 net emissions from the world's largest emitters[12]

Any twenty-first-century strategy needs to place priority on dramatically reducing the per capita and the country's CO_2 emissions to put an end to the code red for humanity and business and to ensure the climate emergency does not impact the planet. The news of the impending devastation should have inspired world leaders on the front lines to, at the very least, hold carbon dioxide emissions to the 1987–90 levels. In reality, we have seen a 61 percent increase in CO_2 and carbon dioxide concentrations, the latter of which are up by 50 percent. This is the reality and the unacceptable result[13].

How have world leaders and business leaders compromised all future generations?

- **By ignoring science for thirty-five years**
  In the past thirty-five years, scientists have confirmed beyond a reasonable doubt that human action and the burning of fossil fuels is the cause of global warming, air pollution, and the climate emergency / code red for humanity. Instead of ending our dependency on fossil fuels a decade after scientists made this fact abundantly clear, we have significantly increased our use of fossil fuels.

- **By increasing the use of pollution-causing fossil fuels**
  This is turn contributed to an increase in CO_2 in the atmosphere from 22 gigatons in the 1980s to 33.9 gigatons in 2021, while concentrations went up from 280 ppm to 420 ppm in 2021.

- **By subsidizing fossil fuels**[14]
  If allowing for the destruction of the planet by overusing fossil fuels is the most catastrophic abdication of global leadership, then the subsidizing of pollution-causing fossil fuels to the tune of US$5.9 trillion per annum in 2021 was nothing short of the compromise of the century as it is this subsidy that is the cause of the code red for humanity that is staring us directly in the face and that will devastate global economies to the tune of US$54 trillion to US$67 trillion in a US$100 trillion economy as we reach the +1.5°C to +2.0°C tipping point by 2025.[15]

There is no doubt that this is the compromise of the century! The lack of progress toward ending our reliance on pollution-causing fossil fuels, itself a result of world leaders and business leaders ignoring the science for thirty-five years, is what has precipitated the crisis. World leaders simply have failed to make any progress to meet the commitments they have made.

The reality is that every scenario being presented is flawed in the first place as there is no guarantee that nations will meet their commitments, as has been recorded in the UNEP Emissions Gap Report in 2021, analyzing the gap between commitments and reality, that is, what nations had, in December 2015, committed to achieve by 2020 as per the NDCs (nationally determined commitments) and what they actually achieved, which was less than 30 percent of what they had promised.

1.1 A Cop-Out at COP26 in Glasgow 2021

It is clear that every scenario to combat climate change looks at our going beyond the +1.5°C tipping point. The range is between +1.8°C and +2.7°C to trigger a catastrophic rise in sea levels[16]. The current crisis where petroleum prices have crossed US$120 per barrel of oil confirms the inaction of most nations, especially those in the European Union, to end their reliance on pollution-causing fossil fuels for the past thirty-five years. We see that as of March 2022, the European Union depends on Russia for 40 percent of its gas. This kind of delay to end our dependency on pollution-causing fossil fuels has caused a crisis, with the free world set to globally ban the use of Russian fossil fuels because of Russia's invasion of Ukraine. Had the EU divested from fossil fuels by 2020, Russia's action would not have been a threat to the EU—or to twenty-first-century-led nations such as Denmark.

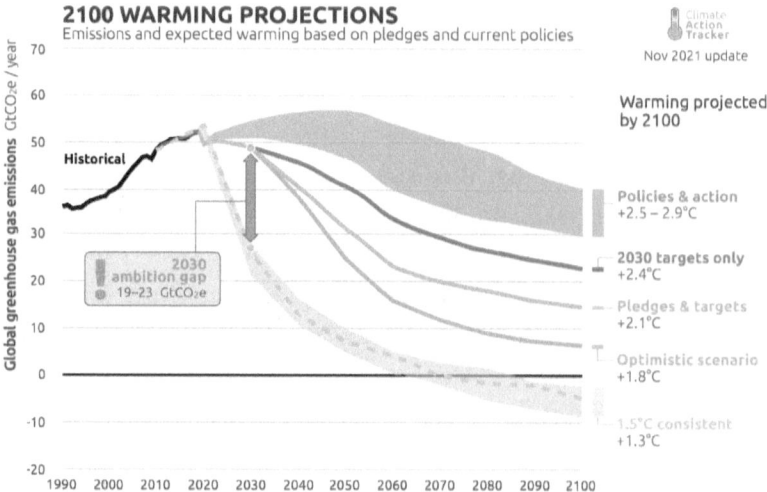

2100 warming projections and post-COP26 warming scenarios[17]

We can either become despondent and give up, seeing as every scenario has us crossing the +1.5°C line, or give climate change our very best shot by implementing a twenty-first-century strategy to fight the code red for humanity and business.

The prognosis is desperate if we leave the fate of our planet to nineteenth-century science-denying leaders who continue along with business as usual. We urgently require twenty-first-century science-led leaders to replace them.

Chapter 2—Twenty-First-Century Implications

1. Nineteenth-century science-denying leaders of nations and businesses without a sustainability mindset have, for thirty-five years, *ignored* all the scientific evidence and the need to end our reliance on pollution-causing fossil fuels and have simply used their influence to postpone doing so from 2000, to 2030, to 2050, and even to 2070!

2. Ninety-two percent of all cumulative GHG/CO_2 has been released into the atmosphere by Northern Hemisphere nations, who should be the ones to fund the energy transition in the Southern Hemisphere.

3. At COP26, China, Russia, India, Saudi Arabia, Mexico, Indonesia, Australia, and Brazil influenced the decision *not* to include any mention of ending their reliance on pollution-causing fossil fuels in the resulting documents.

4. Based on the IPCC and World Meteorological Organization reports, 2025 is the ideal time to have transitioned away from pollution-causing fossil fuels given the 40 percent likelihood that we will cross +1.5°C line by 2025. The latest this should happen is 2030.

5. The subsidizing of fossil fuels to the tune of US$5.9 trillion per annum is the compromise of the century, where nations have subsidized the very cause of the climate emergency.

6. Despite many UN Earth Summit meetings since 1992, the 2015 Paris Agreement, and the COP26 in Scotland in 2021, most nations are 70 percent away from achieving the commitments they made when signing the Paris Agreement as per the UNEP Emissions Gap Report.

7. The current fossil fuel crisis as a result of the Russian invasion of Ukraine, where the European Union depends on Russian gas for 40 percent of its energy requirements, confirms the lack of urgency most nations have felt to end their reliance on pollution-causing fossil fuels.

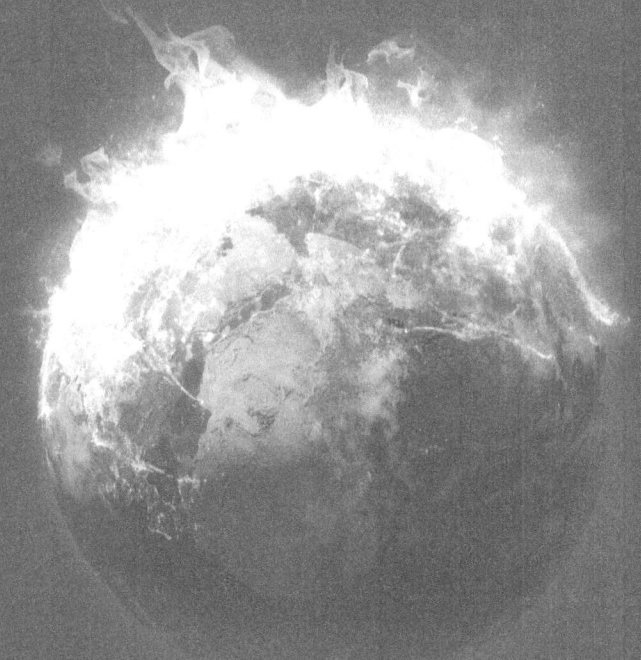

CHAPTER 3

TWENTY-FIRST-CENTURY LEADERSHIP

TWENTY-FIRST-CENTURY LEADERSHIP

To respond to the climate emergency and the code red for humanity and business, we urgently need a cadre of twenty-first-century leaders. The reality is that we have nineteenth-century science-denying leaders who occupy global leadership positions at both the national level and the business level. They have blatantly rejected and ignored the scientific evidence, the impending climate emergency, and the resultant code red for humanity and business.

Business leaders, especially those who work at fossil fuel companies or the banks that fund them, have also actively denied or ignored the science in favor of short-term profits. They have done so to further their own personal agendas or their corporate agendas, to the detriment of the planet. The nations at the forefront of this science denial are those that export fossil fuels: the USA, Russia, Venezuela, Saudi Arabia, and other Gulf nations. Fossil fuel companies such as Saudi Aramco, Exxon Mobile, Gazprom, Adani, Chevron, BP, Shell, Stat Oil, Total, Sinopec, and Petrobras, and all national oil and gas companies, along with banks like JPMorgan, Citigroup, Wells Fargo, Bank America, Barclays, HSBC, Standard Chartered, Sumitomo Bank, Deutsche Bank, and Mitsubishi Bank, have denied the science aggressively.

In addition, the fossil-fuel-driven automobile, airline, shipping, and logistics sectors have used a range of lobbyists—nonscientists—at the expense of the planet for short-term gain. At COP26, nations either exporting or currently dependent on fossil fuels, such as India, China, Russia, Australia, and Saudi Arabia, prioritized profit over the planet. This extremely short-term focus, in turn, enabled and exacerbated the climate emergency, and global warming led to economic devastation caused by extreme weather events. Clearly all these nations and businesses have sacrificed our future generations on the altar of short-term profits.

What we urgently need is a cadre of twenty-first-century leaders who are led by science; are strategic in their decisions, acknowledging the fact that there is neither GDP growth nor quarterly profits on a devastated planet; are sensitive to the fact that ecological devastation delivers a catastrophic blow to humanity and habitats; and know that humanity will be pushed to the brink, where basic needs such as clean air to breathe, fresh water to drink, and fresh food to eat will become impossible to find because, with the 1.5°C to 2.7°C rise in global temperature, with average temperatures of 40°C to 55°C, the planet will simply be inhabitable and impossible to survive on!

3.1 What Twenty-First-Century Leaders Would Have Done

What would have happened after the Bruntland report in 1987 or the recent 2021 COP26 meeting in Glasgow if these events had been attended by twenty-first-century leaders? Instead of ignoring the climate emergency and code red for humanity and business for thirty-five years, those leaders would have felt an urgent need to make an impact on the climate emergency.

World leaders should have unanimously agreed to immediately transition away from pollution-causing fossil fuels used for energy, electricity, and transportation. From 1987 on, the focus would have been on reducing and halting global warming, with every nation, business, household, and individual having begun to end their reliance on fossil fuels in order to eradicate the cause of the climate emergency. Such leaders would have understood that with the code red for humanity and business, every nation and business should have decided to be part of the problem no longer and, instead, become a part of the solution. If such had been the case then, we today would be living on a sustainable planet. The focus and priority would have been on ceasing to use fossil fuels with urgency and embracing renewable energy sources.

Solving every other issue and ignoring the cause of global warming and extreme weather will *not* help us reduce the risks posed by the code red for humanity. Moving to renewable energy, electric vehicles should be the focus. If twenty-first-century science-led world leaders had been in power, they would have done the following:

1) **Stopped subsidizing fossil fuels**—coal, petroleum, diesel, and natural gas—by 2025 or 2030 at the very latest if they had understood the Bruntland, IPCC, and WMO reports and warnings. Instead, in 2021 we invested US$5.9 trillion in fossil fuels.

2) **Transitioned away from using fossil fuels**, having set a strategy for the nation and businesses to do so with urgency. The USA, Europe, China, India, Australia, Saudi Arabia, Mexico, and Turkey would have been renewable-energy-driven by 2030.

3) **Supported fossil-fuel-based energy companies to transition to renewable energy companies.**

4) **Moved to 100 percent renewable energy** (solar, wind, wave, geothermal, and nuclear plus battery storage) by 2030, having created an investment strategy to achieve a minimum of 60 percent renewable energy/electricity by 2025 and 80 percent to 100 percent by 2030.

None of the above has happened in the past thirty-five years, as we've had nineteenth-century science-denying leaders in key leadership positions.

3.2 Twenty-First-Century Leadership Traits

If this is the reality we need to overcome, then we should know the defining characteristics of a twenty-first-century leader. Following are the key traits any twenty-first-century leader must cultivate and be known for. Every twenty-first-century leader will be committed to the four traits below:

- an appreciation for mobilizing science as part of the business strategy
- an urgent need to make an impact on the code red for business
- a sustainability mindset, prioritizing the planet and people over profit
- a willingness to embed the Twenty-First-Century Board Leadership Model in the business's board/C-suite/operations agenda.

As we look at the world today, we see exemplary twenty-first-century leaders who have demonstrated that they are committed to developing all the above-mentioned traits. We find world-class businesspeople who exemplify twenty-first-century leadership leading Google, Tesla, Patagonia, Ørsted, the European Investment Bank, Santander, and Unilever. (How these particular leaders demonstrate the aforementioned four traits in their leadership approach is catalogued in the mini case studies in chapter 4.) These people are global influencers who have relentlessly focused on the need for the leaders of nations and businesses to have a twenty-first-century mindset. Among such influencers, Greta Thunberg and Sir David Attenborough stand out for their consistent and focused message to all world leaders to end their reliance on pollution-causing fossil fuels.

Twenty-First-Century Leadership Role Models

A few twenty-first-century leaders have shown exemplary skills and possess all four leadership traits. I have developed mini case studies on these individuals to inspire all future twenty-first-century leaders. The following twenty-first-century leaders have made an impact on the climate emergency, the health and social emergency, the economic emergency, and the three disruptors—technology, geopolitics, and governance—and by doing so have demonstrated their commitment to the four traits:

- Henrik Poulsen of Ørsted
- Jay Weatherill of South Australia
- Elon Musk of Tesla, Space X, and Solar City
- Sundar Pichai of Google/Alphabet
- Yvon Chouinard of Patagonia
- Paul Polman of Unilever and its Sustainable Living Plan
- Merrill J. Fernando of Dilmah
- Anna Botin of Santander
- Werner Hoyer of the European Investment Bank.

The purpose of mentioning these people is to illustrate the fact that there are some twenty-first-century, science-led, sustainability-mindset leaders who are changing the world as they address both the emergencies and disruptors with great urgency, resolve, and commitment to ensure the organizations they lead are part of the solution to the climate emergency and part of the end to the code red for humanity and business. In every single instance, the four traits that distinguish a twenty-first-century leader are embedded in the individual's behavior and decision-making.

We also feature the following three short reports developed on the following key initiatives from information in the public domain to accentuate how each of them have impact on geopolitics, precision agriculture, and the reality of the compromising banking industry.

1. China's OBOR (One Belt One Road)
2. Wageningen, the Silicon Valley of precision agriculture
3. the report "Banking on Climate Chaos 2021"

3.3 The Twenty-First-Century Leader—Four Defining Traits

There are two make-or-break characteristics that define a twenty-first-century leader:

1) an appreciation for mobilizing science as part of the business strategy
2) an urgent need to make impact on the code red for business (climate emergency).

As you will appreciate, one cannot display the second trait unless one is committed to science and understands the ramifications in the first place. The reason the world has ignored the impending climate emergency for thirty-five years, since 1987, could be that leaders of nations, businesses, and any organization with a global influence have failed to understand or appreciate science adequately enough to respond to the implications of global warming with decisiveness, urgency, and a strategy of zero compromise.

Having an appreciation for mobilizing science as part of the business strategy will lead to respecting the findings and advice of scientists and not ignoring any of the scientific evidence. This in turn should lead to establishing science-based targets so as to impact one's sphere of influence.

An urgent need to make an impact on the code red for business (climate emergency) is ignited only if one understands the ramifications and projected effects and realizes the urgent need to be part of the solution to the code red for humanity.

What we have are nineteenth-century science-denying leaders who have no imperative to be leading nations and thereby compromising the entire planet and its people with one the following responses:

- simply ignoring the code red for business and the climate emergency
- taking cosmetic actions to imply the climate emergency is being tackled simply by managing the optics

- taking insufficient action in areas that have no effect on the main problem
- communicating that any actions to be taken will be taken in the long term, that is, by 2040 or 2070, and offering convincing arguments so as to avoid being scrutinized while climate change devastates the planet.

All the foregoing are actions taken by nineteenth-century science-denying nonleaders who have either not passed Science 101 or have decided what matters is short-term results. This is the dilemma every leader is confronted with. It is increasingly clear that the behavior shown after COP26 behavior is the rule and not the exception for what happened after previous COP meetings and every other single global forum thus far, ending in short-termism and resulting in fulfilling the economic self-interests of science-denying leaders put profits ahead of the planet and people. So, how do you think that you could be a twenty-first-century leader and embed the top two traits in your thinking, attitude, and behavior?

An Appreciation for Mobilizing Science as Part of the Business Strategy

We don't need leaders to be scientists, just people who have an appreciation and uncompromising respect for scientific thinking. Over the centuries, science has been responsible for the rapid advancement of humanity with the past sixty years having seen groundbreaking technologies such as nanotechnology, artificial intelligence, robotics, precision agriculture, 3D printing, nano–bio, and blockchain. These, along with unprecedented advances in medicine, space exploration, and increasing the human life span, prove that science has been the single-most common force for global advancement. How is it that leaders of nations and businesses ignored and overwhelmed every single warning, trend forecast, prediction, and potential risk to the planet that was ever presented and persisted with pollution-causing fossil fuels and the corrupt lobbying of the fossil fuel industry for so long? The *only* answer is that they were either ignorant of science or corrupt to the core—or both.

A twenty-first-century leader will engage the best scientific advisers and populate his or her national advisory team or corporate board with world-class scientists who do not compromise scientific knowledge or hypotheses gleaned from years of observation and experiment, proverbially selling their souls for a pot of gold. Leaders need to take the time and expend the energy to understand the science of the climate emergency, isolate the direct and indirect effects that nations and businesses have on Scope 1, 2, and 3 emissions, and aggressively reduce these effects in order to move urgently from being a part of the pollution problem to being part of the pollution solution as they respond to the code red for humanity and business.

Every leader who has been given stewardship of a nation or business and therefore has the authority to affect everything within his or her entire sphere of influence should take it as a serious responsibility to make the transition away from pollution-causing fossil fuels and also make strategic decisions to ensure that future generations will be able to survive on this planet. Today, the UN makes compromises, communicating the current reality in such a way so as not to offend any of the key nations that contribute to its budget, ensuring that both the IPCC and the WMO reports cause no alarm, and setting deadlines far into the future, such as in 2050 to 2070, instead of issuing an urgent ultimatum to end the world's reliance on fossil fuels by 2025 or 2030.

An Urgency to Impact the Code Red for Humanity (Climate Emergency)

How can any leader who denies science ever feel an urgent need to make an impact on the climate emergency? We need a cadre of twenty-first-century leaders who understand that there is no business to be done on a planet ravaged by a climate change, one where fresh water, fresh air, and fresh food will be at a premium. This knowledge should ensure that a leader will act quickly to develop the third and fourth traits, as follows:

- a sustainability mindset, prioritizing the planet and people over profit
- a willingness to embed the twenty-first-century board leadership model in the business's board/C-suite/operations agenda.

As previously mentioned, the pivotal, all-influencing trait that defines a twenty-first-century leader is an appreciation for mobilizing science as part of the business strategy.

With the third trait, we have a litmus test to which a twenty-first-century leader may subject every single decision and investment:

a) Does the investment proposal compromise the planet or people in any way to help generate profits or GDP growth?
b) What are the optional twenty-first-century strategies we can consider to achieve the same end goal without either compromising the regenerative process or harming the environment or humanity?
c) How can we minimize the negative impact on the planet and people?

Developing trait 4 is possible only if one first knows the answers to the foregoing three questions and thereafter populates the board, C-suite, operations, and management agendas of one's organization with modules from the Twenty-First-Century Board Leadership Model, not only providing the key response strategies in the agenda but also leading the implementation of such strategies with commitment and a drive to inspire the entire organization to act with urgency and impact the twenty-first-century strategy.

The fourth trait is what one leverages to build a twenty-first-century strategy to respond to the emergencies and disruptors and, more importantly, to execute that strategy with urgency and commitment. The key differentiator of a twenty-first-century leader, therefore, is that he or she considers *living the four strategic traits* and acting on them with urgency as a prerequisite to leading his or her organization. The key is that such a leader recognizes that traits 1 and 2 are nonnegotiable and demand a response and that the way to provide this response is to first consider traits 3 and 4, where the leader's urgency is demonstrated by setting aggressive and focused targets.

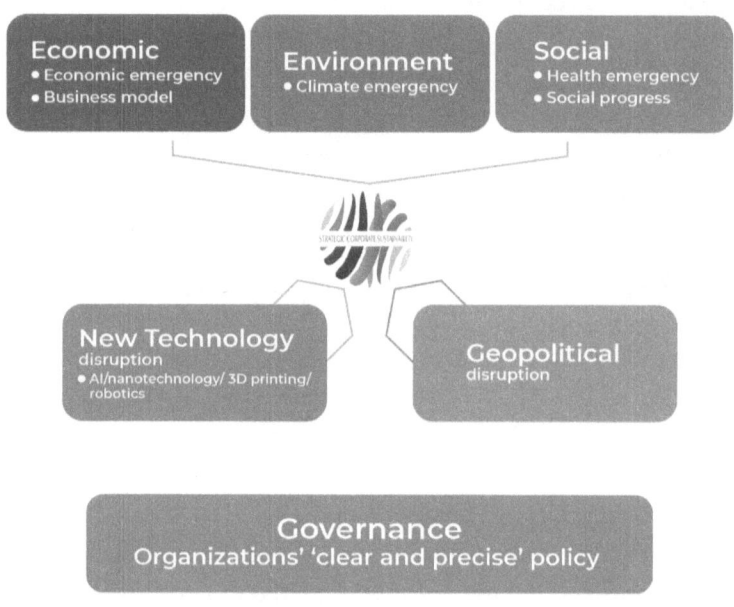

Twenty-First-Century Board Leadership Model©[1]

Creating the Twenty-First-Century Board Leadership Model

The Twenty-First-Century Board Leadership Model© was conceptualized and created in June 2020.

Many factors inspired its creation, including the concept of a triple-bottom-line sustainability mindset, which was created by John Elkington in 1997[2]. The need for twenty-first-century leaders to be people who prioritize the planet and people over profit is an essential component of the model. Businesses must build business models that deliver economic growth and profit sustainably, ensuring that every decision that affects the planet or the environment is analyzed and evaluated before they determine a course of action that will make the least negative impact and leave behind the smallest carbon footprint while contributing to the social progress of the employees and the community.

Another thing that influenced me to create the model was Yuval Noah Harari's *21 Lessons for the 21st Century* (2018), which, as the title implies, articulates twenty-one lessons for the twenty-first century. These relate to the key challenges that every single twenty-first-century leader of a business must consider as he or she is devising twenty-first-century strategies. Challenges such as the climate emergency and technological and geopolitical disruption, to which a majority of business strategy programs pay little attention, become key to developing strategic initiatives to address such things comprehensively.

Finally, the concept of strategic corporate sustainability inspired me to create the Twenty-First-Century Board Leadership Model. It is key that any twenty-first-century strategy must first embed sustainability in the corporate strategy and, second, distinguish the organization by its sustainability paradigm. When we combine the concept of triple bottom line, the twenty-one lessons for the twenty-first century, and strategic corporate sustainability, we arrive at the Twenty-First-Century Board Leadership Model, which combines all three concepts, and the learning is clustered into six modules covering three emergencies and three disruptors.

Any leader not fully apprised of and ready to face the very real twenty-first-century challenges is both irrelevant and incapable of dealing with such challenges. The reality is that there are nineteenth-century science-denying people in positions of power. We need the Twenty-First-Century Leadership Model because a majority of leaders have ignored the reality of climate change and the science-based solutions to this problem.

The Twenty-First-Century Board Leadership Model

All three concepts mentioned earlier, along with relevant strategic issues, are presented in six modules, one for each of the three emergencies (the climate emergency, the health and social emergency, and the economic emergency) and one for each of the three disruptors (technology, geopolitics, and governance).

The six modules present knowledge that is essential to have when making any twenty-first-century decision. The reality is that many of these modules are not included in mainstream curriculums, which is confirmation that such institutions of learning are failing to prepare twenty-first-century leaders to face the world's twenty-first-century challenges. The model is not a substitute for the curriculum being covered in mainstream institutions but is an essential supplement to what is being taught to future twenty-first-century leaders.

Following are three unique features that clearly set the Twenty-First-Century Board Leadership Model apart with its dynamic, relevant content:

- **The Twenty-First-Century Strategy Template©**, a game-changing feature offering every person reading *Twenty-First-Century Leadership to Fight the Code Red for Business* the opportunity to create a company- or organization-specific twenty-first-century strategy that can then be presented to the board or C-suite for deliberation, fine-tuning, and adaptation to the corporate, national, or organizational strategy.
- **The Twenty-First-Century Board Leadership Model MasterClass©**, which is currently delivered as a six-module series. The dynamic nature of the content ensures that much of what is relevant and pertinent today is not expected to be relevant in the future. A key feature is that each module is interconnected with

the others, with all of them (i.e., the three emergencies and three disruptors) being interdependent and dynamic. Twenty-first-century leaders are sensitized to address each of the challenges presented in each module in a global, 360° sense, as opposed to addressing issues in isolation, accepting the fact that we operate in a very dynamic environment that requires strategic decisions that take into account the prevailing environment at the time.

- **A dynamic and living curriculum.** The Twenty-First-Century Board Leadership Model MasterClass curriculum is constantly updated by a dedicated team of world-class researchers to challenge each participant to create a strategy that is current, relevant, and strategic, thereby showing an appreciation for the interconnectedness of all six modules.

The Twenty-First-Century Board Leadership Model MasterClass is not a substitute for an MBA, an executive education program, or an Institute of Directors governance program. It is a dynamic, living supplement for anyone seeking to be a relevant twenty-first-century leader.

The advice of Raymond Schadeck, a former chairman of ILA, played a key role in my decision to create an interactive MasterClass using a truly unique world-class model. We are proud to have inspired a cadre of more than one hundred world-class twenty-first-century leaders in a relatively short period of time, where the crème de la crème of business leaders were inspired to create a Twenty-First-Century Strategy Template for their businesses.

Twenty-First-Century Leadership 2025

It is our vision to engage the unique new cadre of twenty-first-century leaders who have either read *Twenty-First-Century Leadership to Fight the Code Red for Business* or followed the corresponding MasterClass in any nation of the world. Following are ways to engage with the content:

- By reading the book *Twenty-First-Century Leadership to Fight the Code Red for Business*, to be launch in 2023.

- **By attending the Twenty-First-Century Board Leadership Model MasterClass**, which will continue to build a cadre of world-class twenty-first-century leaders in Europe and Asia.
- **By following the Twenty-First-Century Global Leadership group on LinkedIn**, launched to create a community of all those who have completed the MasterClass so they may be united and encourage each other.
- **By subscribing to the Twenty-First-Century Strategy YouTube channel**, featuring all key lectures given on any of the modules in 2022.
- **By attending the Twenty-First-Century Global Leadership Forum** by 2025 to join with all those who have participated in the MasterClass and read the book.
  - Showcase the successful implementation and impact of a Twenty-First-Century Strategy Template.
  - Identify the two twenty-first-century challenges that require urgent attention and engagement, then develop twenty-first-century strategy recommendations for implementation by the twenty-first-century leadership community, to be reviewed at the next annual event.

  If you are reading *Twenty-First-Century Leadership to Fight the Code Red for Business*, chances are that you have already completed the MasterClass or are considering enrolling. Be inspired by and proud of the fact that you belong to an exclusive new cadre of twenty-first-century leaders who will redefine strategy and have the responsibility for being strategic stewards of the planet's resources, not simply exploiters of its finite resources for short-term profits.

As you lead your organization or business, what key questions should you be asking, and what strategies should you consider, if you want to embed the Twenty-First-Century Board Leadership Model and create a twenty-first-century-relevant and -ready strategy?

Chapter 3—Strategy and Implications

Twenty-first-century leaders will include the Twenty-First-Century Strategy Template actions in the board agenda and review the impact of and strategies for addressing each of the emergencies and disruptors:

- How do we ensure that we have a CEO or board with a twenty-first-century mindset making decisions to fight the code red for business?
- How does the current business model or supply chain strategy affect the climate emergency, the health and social emergency, and/or the economic emergency? What urgent changes are needed for an organization to be part of the solution to the code red for business?
- What strategic changes are required to minimize the potential negative effects of a technology, geopolitical, or governance disruption on the business model? What are the strategic risks and opportunities?
- What urgent and decisive action needs to be taken to circumvent and eliminate the negative effects on the planet and on people (starting with your own employees, the community your business operates in, and society in general)? How can the company inspire everyone within its sphere of influence to live sustainable lifestyles, not just engage in corporate social responsibility in order to look good?
- In order to prioritize the planet and people over profit in all decisions, a leader must generate a comprehensive set of options and alternative strategies that offer clear direction for deploying a final strategy that makes the least possible impact on the planet and people. We cannot eliminate risk or the impact on the planet, but a twenty-first-century leader will be able to minimize these factors.

This is why it is a prerequisite for all leaders to be up to speed with twenty-first-century science-led thinking and to have sustainability mindset, versus being a nineteenth-century science denier in a position of influence in business or on a board whose intent is to minimize the risk that comes with flawed decisions that promise to make a negative impact on both the planet and people.

CHAPTER 4

THE TWENTY-FIRST-CENTURY BOARD LEADERSHIP MODEL

THE TWENTY-FIRST-CENTURY BOARD LEADERSHIP MODEL

4.1 The Emergencies

As a science-led leader with a sustainability mindset, you must face re- ality and act with urgency. You must create, not a frivolous time line for ending dependence on fossil fuels sometime between 2050 and 2070, but a concrete deadline soon in the future, such as by 2025. A twen- ty-first-century leader has an understanding of the direct and indirect effects of all decisions made pertaining to the business model and the supply chain. The three emergencies need to be addressed with focus, tenacity, and urgency to minimize their impact and ideally eliminate all negative effects so the organization may move toward developing solutions to address all three emergencies—the climate emergency, the health and social emergency, and the economic emergency.

Unfortunately, emergencies are downplayed and action postponed with a "Let's address this issue after we first make all the money," while the company pollutes the planet and enables the emergency. Nineteenth- century nonleaders of nations, businesses, and the fossil fuel and banking industries have denied the science. By continuing to burn fossil fuels, they have significantly compromised the lives of future generations by causing the temperature of the planet to rise, engendering the climate emergency as we near the +1.5°C mark, which will occur sometime between 2025 and 2040. This is turn will set off the catastrophic destruction of the planet as the Arctic, Antarctic, and Himalayan ice caps melt, enabling an unprecedented rise in sea level, destroying the reflective white sheet, dramatically reducing the global freshwater stock, and contributing to putting the Earth's orbit off-balance. Have world leaders become so obsessed by short-term profits generated by fossil fuels that they are happy to ignore the long-term emergency, a term used by the medical profession to signal the utmost urgency to act immediately? Do we move patients to the intensive care unit for treatment thirty to fifty years after the fact? Do we call the fire department thirty or fifty years after a raging fire starts, asking them to come and put it out? This is why we need twenty-first-century leaders to act on the three emergencies, with urgency, by 2025.

4.2 The Climate Emergency Module

In the climate emergency module, the first and second traits of twenty-first-century leadership are demonstrated by the vision, the decisions made, the investments made, and most of all the pace at which transformational change is achieved. As a reminder, the first and second traits are as follows:

1) an appreciation for mobilizing science as part of the business strategy
2) an urgent need to make an impact on the code red for business (climate emergency).

Nineteenth-century science-denying leaders of nations and businesses can no longer be allowed to compromise all future generations. We need to hold them accountable and replace them with twenty-first-century leaders ASAP to give all future generations a chance of survival on a sustainable planet. The fact that we have allowed our current leaders to drastically affect all future generations is clearly highlighted in the World Economic Forum (WEF) Global Risks Report 2022[1] as climate action failure.

The key focus of the climate emergency module is to inspire businesses to do the following:

4.2.1 Transition Away from Pollution-Causing Fossil Fuels (Coal, Petroleum, Diesel, and Natural Gas) to Reduce the Carbon Footprint

The priority is to reduce and minimize the carbon emissions of businesses that are a result of a business's many activities. Carbon emissions are the direct cause of global warming, the climate emergency, and the code red for humanity and business. Emissions need to be reduced systematically after establishing the carbon footprint for Scope 1, 2, and 3 emissions and targeting the end of the use of fossil fuels and the transition to renewable energy, sustainable transportation, and sustainable options.

4.2.2 Create Sustainable Supply Chains by moving from outsourcing and offshoring to near-shoring and insourcing to reduce the carbon footprint.

We need to address the following questions as we respond to the climate emergency:

- What is the carbon footprint (Scope 1, 2, and 3 emissions) of the organization (annual appraisal)?
- What is the energy and electricity component of that which directly impacts global warming?
- How can we reduce the carbon footprint and move to 100 percent renewal electricity/energy and sustainable transport by 2025 or 2030 at the latest?

The World Meteorological Organization and IPCC forecast that the earth's temperature will rise by 1.5°C between 2025 and 2040. Any commitments one makes to end reliance on pollution-causing fossil fuels after that time will not be part of the solution to the climate emergency or the code red for business.

4.3 The Inspirational Energy Transitions
of Ørsted and South Australia

The case of Ørsted in Denmark, where twenty-first-century leader Henrik Poulsen led the transformation of the Danish Oil and Gas Company to become among the world's leading wind power companies, is inspirational. The twenty-first-century leadership of South Australia, where despite a nineteenth-century science-denying prime minister who has been advocating persisting with pollution-causing coal, has led South Australia to be 100 percent renewable energy driven for electricity and has been moving rapidly every year since 2017 to make the country more sustainable in every way. These two cases prove that any energy company or nation dependent on pollution-causing fossil fuel that wants to end its reliance can achieve its energy goals.

The Ørsted and South Australia cases prove beyond a reasonable doubt that all it takes is a twenty-first-century, science-led, leader with a sustainability and a board with a strategic twenty-first-century vision to make a revolutionary change from being dependent on fossil fuels to ceasing to use them and moving to renewable energy.

These are two inspirational cases that challenge every fossil fuel company to consider and create a strategic plan to move from a pollution-causing fossil fuel company to a 100 percent renewable energy company by 2025 at the latest. These cases also confirm that every nation, region, and city that is considering moving to 100 percent renewable energy, irrespective of being led by nineteenth-century science-denying leaders and being bound by their policies that support fossil fuels such as coal and natural gas, can end its dependency on pollution-causing fossil fuels and move to a profitable and efficient renewable energy strategy.

The energy transition needs the direct intervention of government to set terms and create policies and incentives to help all the fossil-fuel-driven energy companies, both public and private, remain energy companies but transform to renewable energy companies. Ideally, the incentives should be in the form of capital, sustainable financing, and tax incentives.

Reluctance to do so by nations with major fossil fuel reserves, such as Saudi Arabia, Russia, Australia, Indonesia, Poland, and Venezuela, is to be expected. The simple case for renewable energy, demonstrating that the cost of fossil fuel extraction versus harnessing sun and wind power with battery storage, has to be made comprehensively. However, the cost of infrastructure already invested in will inhibit the transition. This is the challenge Danish Oil and Gas had to overcome as its twenty-first-century leader Henrik Poulsen led the transformation to Ørsted.

The Ørsted case is clearly the benchmark case for twenty-first-century leadership and energy transition from pollution-causing fossil fuels to renewable energy. It could not have happened if the CEO and board of directors had not had a twenty-first-century mindset and, no doubt, the vision and the commitment of the people of Denmark to follow through.

4.3.1 Henrik Poulsen, Former CEO of Ørsted, Leading the Energy Transition[2][3][4][5][6][7]

Ørsted (formerly Danish Oil and Natural Gas—DONG Energy) is a renewable-energy-focused company based in Denmark with a revenue of €7.1 billion (US$8 billion) in 2020. It employs around 6,672 people in Denmark, Sweden, the United Kingdom, Germany, the Netherlands, the United States, and Taiwan.

Ørsted constructs and operates offshore and onshore wind farms, solar farms, energy storage facilities, and waste-to-energy facilities for its customers. It is recognized by the CDP's Climate Change A List as a global leader on climate action. The Corporate Knights has ranked Ørsted the world's most sustainable energy company in the Global 100 Index for the past three years (2019–22). The company was placed fourth in the global ranking of the top one hundred companies for sustainability in 2022 according to an independent assessment carried out by *Sustainability* magazine.

With the twenty-first-century science-led leadership of Henrik Poulsen (former CEO of Ørsted), the company transformed the fossil fuel energy business into a renewable energy company between 2009 and 2019 by aggressively investing in renewables, in particular, offshore wind. Danish Oil and Natural Gas transitioned from 85 percent fossil fuels in 2009 to 75 percent renewables in 2019 with the world's largest offshore wind farm (one thousand offshore wind turbines, creating 3.9 gigawatts of power, which can power ten million homes). In 2020, 90 percent of total energy generation was from renewable sources. In 2021, the European Investment Bank (EIB) signed a loan for five hundred million euros with Ørsted to support the latter's renewable energy projects.

Under the current CEO, Mads Nipper, the vision of the company is to become the world's leading green energy provider by 2030 by doing the following things:

- ✓ transitioning entirely away from coal by 2023
- ✓ being carbon-neutral by 2025 for its energy generation and company operations
- ✓ investing US$57 billion in renewables by 2027
- ✓ installing 11–12 gigawatts of total offshore wind capacity by 2025 (in 2020, the company had reached 7.6 gigawatts)
- ✓ installing a total of 30 gigawatts of offshore wind capacity by 2030 (in 2020, the company had reached 7.6 gigawatts)
- ✓ installing 50 gigawatts of renewable energy capacity by 2030.

Henrik Poulsen of Ørsted has demonstrated to the world the following twenty-first-century science-led leadership traits:

- He has shown an appreciation for mobilizing science as part of the business strategy by leading the transition from a fossil-fuel-based company to a company driven by renewable energy.
- He has shown an urgent need to make an impact on the code red for business as all his climate targets are set to be reached before 2030.
- He has shown evidence of a triple-bottom-line mindset, prioritizing the planet over profit.
- He has shown a willingness to embed the Twenty-First-Century Board Leadership Model in his business's board/C-suite/operations agenda—a transformation from 2009.

4.3.2 Jay Weatherill (Former Premier of South Australia)—Leading South Australia's Energy Transition[8][9][10][11][12][13]

For the second case of energy transition, we look to South Australia, where we see the visionary leadership of the government going against the policies of a nineteenth-century coal-power-advocating prime minister. The ability to challenge the status quo with conviction based on science is clearly demonstrated here.

South Australia is home to around 1.7 million people and is almost one and a half times the size of Texas. The South Australia state government has made a stunning transformation to green energy in the past fifteen years, with renewables moving from 0 percent in 2006 to meeting 60 percent of South Australia's electricity needs by 2020. Key to the state's success was a commitment to a diverse mix of large-scale renewable energy generation and storage, such as wind, solar thermal, solar PV, bioenergy, battery, pumped hydro, and thermal storage.

Other notable achievements for 2020 include having the world's largest per capita rollout of home battery storage; wind farms accounting for 40 percent of Australia's wind energy; and one in three homes with rooftop solar. The state is also on track to become 100 percent driven by renewable electricity by 2030, with twenty billion dollars of renewable projects in the pipeline, and net zero for energy by 2050.

This radical shift was facilitated by a twenty-first-century science-led leaders with a sustainability mindset, Jay Weatherill, the former premier of South Australia, who served from October 21, 2011, to March 19, 2018. Despite a lack of support from Prime Minister Scott Morrison and the federal government, Weatherill forecasted that aggressively shifting to renewables and responding to the code red for humanity and business would unlock innovation and economic opportunity while making South Australia a more livable, smarter, more resilient state.

Jay Weatherill has demonstrated to the world the following twenty-first-century science-led leadership traits:

- an appreciation for mobilizing science as part of the business strategy, as evidenced by his rejection of coal power and his transition to a diverse mix of large-scale renewable energy generation and storage

- an urgent need to make an impact the code red for business by 2025, moving from 0 percent renewables in 2006 to meeting 60 percent of South Australia's electricity needs by 2020.
- a triple-bottom-line mindset, moving toward electric transportation
- a willingness to embed the Twenty-First-Century Board Leadership Model in his business's board/C-suite/operations agenda, as shown by the results achieved between 2011 and 2018.

The two aforementioned cases demonstrate beyond a reasonable doubt that if you have twenty-first-century science-led leadership at the board level and in the C-suite, then you can make the energy transition to 100 percent renewables. It is significant to mention that by 2022, the cost of renewable energy has now achieved parity with, and sometimes is even lower than the cost of, pollution-causing fossil fuels.

At Ørsted, we see the twenty-first-century leadership of Henrik Poulsen, who made a strategic and bold decision to ensure that the Danish Oil and Gas Company would be transformed to be part of the solution as a wind power giant fighting the climate emergency, versus being a key part of the problem. Can all fossil fuel companies transform themselves to be part of the solution? Yes, provided they have twenty-first-century science-led leaders with a sustainability mindset at the helm. Can such companies be profitable and reinvent themselves? Yes. Ørsted is both profitable and viable as one of today's renewable energy giants. Let this be a lesson to Saudi Aramco, Exxon Mobil, Total, Chevron, Gazprom, BP, and Shell that they can remain energy solution providers without being part of the pollution problem.

Jay Weatherill, former premier of South Australia, led his country's energy transition. His twenty-first-century leadership, standing up to the nineteenth-century science-denying Prime Minister Morrison. Morrison has dramatically compromised the planet, including Australia, by insisting his country will pursue coal power and by supporting the Adani coal mine in Australia despite devastating wildfires, floods, and extreme weather, leading to destruction of the Australian economy. Weatherill transformed South Australia to become 100 percent renewable energy driven and took action to create an electric vehicle charging network to enable EVs to thrive in South Australia.

It's clear to see all four key traits of twenty-first-century leadership in both Poulsen and Weatherill, both of whom are science-led and seek science-driven solutions. They both felt an urgent need to make an impact on the climate emergency, with Ørsted and Weatherill addressing the key energy transition in order to be part of the solution by 2022. Twenty-first-century leaders know that moving to renewable energy is the key to fighting the code red for humanity and business.

Renewable energy, it should be noted at this point, is energy that is collected from renewable sources that are naturally replenished on a human timescale. We need to solve the energy crisis to halt global warming, and the transition away from all pollution-causing fossil fuels—coal, petroleum, diesel, and natural gas—is the key.

The twenty-first-century science-led leader with a sustainability mindset knows that the urgent move to renewable energy is the key and the focus of UN sustainable development goal number seven: renewable energy sources. Moving to renewable energy sources is the key to this urgently needed action. Pursuit of the following energy sources is the way forward:

- advanced perovskite solar panels plus battery storage (these could be roof or ground mounted, or mounted on bodies of water (floating solar)
- wind power plus battery storage
- tidal wave energy plus battery storage
- geothermal energy
- hydroelectricity
- nuclear energy.

These are the key options available for us to transition away from pollution-causing fossil fuels, which cause global warming as a result of the accumulation of greenhouse gases / carbon dioxide (CO_2).

Renewable Energy Driven

Copyright© by R. A. Fernando - Design by Design Nariya

4.4 Climate Emergency and Supply Chains

Every supply chain is vulnerable to the health and social emergency and the climate emergency. No supply chain on our globally connected planet is climate-, health-, and social-emergency-proof. They are all directly vulnerable to every extreme weather event and the social effects of the health emergency. As human beings yet play a key role in the supply chain with regard to both manufacturing and logistics, until such a time as robotics, automation, AI, and precision agriculture solutions replace human labor, there is no solution in the short to medium term to cope with the climate emergency and extreme weather incidents. This confirms the need to create twenty-first-century supply chains.

4.4.1 Create Twenty-First-Century Supply Chains

The second focus of the climate emergency module is to reduce the carbon footprint of all supply chains by moving from outsourcing and offshoring to near-shoring and insourcing. In 2022, we see that almost every supply chain is being disrupted either by issues related to COVID-19 and its variants or by extreme weather incidents.

It may already be too late to put an end to the devastation of every supply chain. A strategic response is required to minimize the negative effects of the crisis.

In the nineteenth century, the outsourcing of supply chains was considered the right thing to do, and most nations allocated their sourcing of raw materials, processing, manufacturing, and logistics to a selected supply chain partner. China became the factory of the world and, because of its strategic investment in technology, its manufacturing capability, its mobilization of manpower, and its efficiency and scale, was unbeatable in terms of unit costs. However, even China with all its aforementioned strengths and insurmountable advantages was directly impacted by both COVID-19 (health emergency) and extreme weather (climate emergency).

4.5 The Health and Social Emergency Module

Gaining an understanding of the health and social emergency is possible only if one has an appreciation of science and the many negative things created by businesses and humankind that cause health emergencies and social inequality, most times by ignoring the science and the reality of social injustice.

The focus of this module is to ensure all business leaders commit to developing an appreciation for science and develop a strategic response to the climate emergency.

4.5.1 The Health Emergency—a Code Red Strategy

We need to anticipate more pandemics as we destroy forest cover around the world and more animals are caught up in the wet markets of Asia. SARS, H1N1, bird flu, Ebola, and COVID-19 all support this point. COVID-19 is just the beginning of many global pandemics that will force nations to regularly impose lockdowns on their economies and economic activity, causing air travel and supply chain logistics to come to a halt as we saw happen in 2019, 2020, and 2021.

There are many strategic questions about operating in a world that will be impacted by future pandemics and health emergencies that twenty-first-century business leaders must answer. How does a business operate its manufacturing department with a limited number of employees? Can the business automate and introduce robotics and AI-driven manufacturing in the future? Do employees need to be physically present in an office to get work done? Would it not encourage greater productivity and efficiency to allow employees to work from home two or three days a week? How does a business shift distribution directly to consumers? How does it build connectivity and inspire the workforce using the Internet of Things (IoT).

These are all topics that need to be dealt with to ensure business operations continue with strategies ready to be implemented within twenty-four hours. Developing a code red strategy is an exercise a business must engage in to ensure it has an off-the-shelf game plan ready to be mobilized in twenty-four hours the next time a pandemic arises. The strategy should address how a service-based business can remain operationally efficient even if all its employees are unable to get to work. If it's a manufacturing organization, then it needs to develop a minimum workforce strategy to deliver a base volume of products.

4.5.2 COVID-19 and Pandemics—Building a Code Red Strategy

How did nineteenth-century science-denying leaders handle the COVID-19 health emergency? In 2019–20, many science-denying leaders in the USA and Brazil ignored the science and the medical advice as they had done with the climate emergency, also ignoring the need to transition away from using pollution-causing fossil fuels. They all prioritized profit over the planet. As per a Johns Hopkins report, these nations account for the highest number of fatalities year to date as they rejected all medical and health advice at the start.

4.5.3 Key Health Emergency Issues—the Linkage

When a company addresses the health emergency, it needs to review how each of the following issues are interconnected and how they affect the business directly and indirectly: deforestation, forest fires, and water scarcity.

The rain cycle is directly impacted by deforestation. Global water stress is exacerbated by deforestation. If the company is enabling deforestation to grow crops such as palm trees (for palm oil) and soybeans, then it is directly complicit in deforestation. If the company maintains a dependence on paper, versus being paperless, in this day and age, once again it is enabling deforestation.

The lack of proactive action by science-denying nineteenth-century leaders in Australia, Russia, and the USA and the wanton destruction of the lungs of the earth—the Amazon rainforest—by Brazil's nineteenth-century science-denying leader are two key facts that confirm the danger of having science-denying leaders running nations or businesses. Global warming, caused by the burning of fossil fuels, has resulted in extreme weather events with unprecedented temperatures leading to droughts and forest fires.

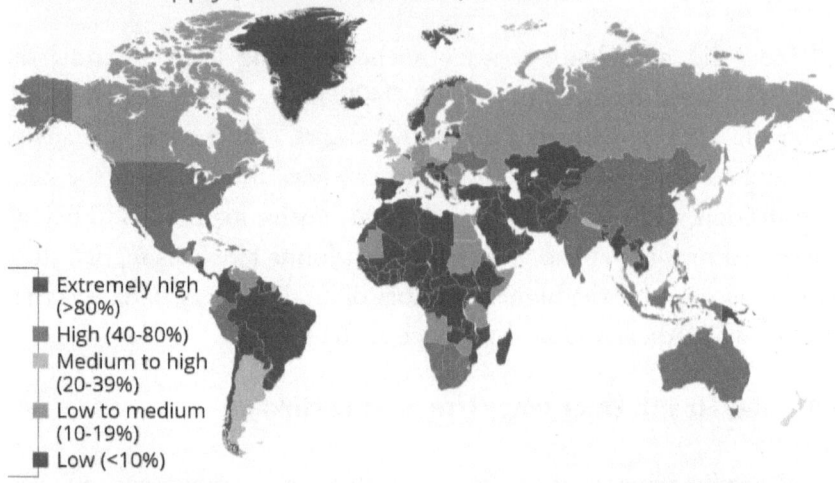

Where Water Stress Will Be Highest by 2040

Projected ratio of water withdrawals to water supply (water stress level) in 2040

- Extremely high (>80%)
- High (40-80%)
- Medium to high (20-39%)
- Low to medium (10-19%)
- Low (<10%)

Source: World Resources Institute via The Economist Intelligence Unit

statista

Where will water stress be highest in 2040?[14]

How are nations balancing economic growth with environmental goals? The INSEAD case study covering Singapore, "a city in a garden," demonstrates that twenty-first-century leadership can balance the two.

4.5.4 Singapore's Former Prime Minister Lee Kwan Yew—Triple-Bottom-Line Mindset [15] [16] [17] [18] [19] [20]

Global Model for Water and Waste Management (Health and Social Emergency)

Singapore is one of the most densely populated countries in the world, housing 5.7 million people in an area of 728.6 km², with four biodiversity-rich nature reserves covering around 33.26 km² (2020).

The country has seen steady economic growth with an annual GDP growth rate of 6.23 percent from 1976 to 2021. The World Intellectual Property Organization (WIPO) has ranked Singapore eighth in the Global Innovation Index 2021 rankings.

This clean, green city allows for a high quality of life, the foundation for which was laid by the twenty-first-century, science-led, sustainability-mindset leadership of former prime minister Lee Kwan Yew (1959–90). He led from the front, spearheading a one-week drive in 1969 to clean up Singapore with thousands of volunteers. He also attracted foreign investments and introduced good governance in the form of systems integration, infrastructure, processes, and integrated long-term planning for green spaces, water, and waste management.

Over the last decade, Singapore has shown signs of being highly vulnerable to extreme weather events, with the mean temperature increasing from 26.9°C to 28°C from 1980 to 2020, rainfall increasing from 1.2 mm to 1.7 mm from 1975 to 2009, and 30 percent of land being only five meters above sea level.

To make a positive impact on the climate emergency, the current prime minister, Lee Hsien Loong (son of Lee Kuan Yew), has unveiled the Singapore Green Plan 2030. The plan is a whole-nation movement to advance Singapore's national agenda on sustainable development. The ten-year plan is overseen by five of the main ministries: the Ministry of Education, the Ministry of National Development, the Ministry of Sustainability and the Environment, the Ministry of Trade and Industry, and the Ministry of Transport. The pillars covered in the plan include City in Nature, Energy Reset, Sustainable Living, Green Economy, and Resilient Future. Some notable targets in the plan include the following:

- ✓ By 2030, 50 percent more land—around two hundred hectares—for nature parks (every household will be located within a ten-minute walk of a park).
- ✓ By 2030, an aim to reduce the waste sent to Singapore's landfill by 30 percent (new and recycled construction material made from waste ash).

- ✓ By 2030, a goal to achieve a 70 percent overall recycling rate: 81 percent nondomestic, 30 percent domestic, to help extend the Semakau Landfill's life span beyond 2035.
- ✓ By 2030, a goal to increase EV charging point targets from twenty-eight thousand to sixty thousand.

Water is inextricably linked to the climate emergency. For more than two decades, Singapore's national water agency, PUB, has been successfully driving the Four National Taps strategy (water from local catchment, imported water, NEWater [benchmark for closed-loop water treatment], and desalinated water).

Despite the lack of groundwater resources in the country and Singapore's reliance on Malaysia for imported water, the PUB has crafted a plan for Singapore to become largely self-sufficient in terms of water by 2060 (when the population is projected to double) by doing the following:

- ✓ Increasing the current 40 percent share of NEWater (of the total water supply) to 55 percent by 2060.
- ✓ Developing desalination treatment plants to supply 30 percent of the nation's water needs by 2060.
- ✓ Relying on local catchments (utilizing a maximum amount of area for rainwater harvesting), which include seventeen reservoirs, with only a small percentage of imported water, for the remaining share of 15 percent.

Singapore's water and waste management has become a global model for mitigation of the climate emergency. The country has demonstrated to the world that with good governance and innovation, economic development can take place without comprising the environment.

Singapore's leadership has demonstrated to the world the following twenty-first-century science-led leadership traits:

- • an appreciation for mobilizing science as part of the business strategy, by demonstrating a triple-bottom-line mindset in the 1960s

- an urgent need to make an impact on the code red for business, by created a prosperous garden city (30 percent tree cover)
- a triple-bottom-line mindset, by creating the multiministry-run Singapore Green Plan 2030
- a willingness to embed the Twenty-First-Century Board Leadership Model in the business's board/C-suite/operations agenda, prioritizing the planet and people over profit.

The Singapore case demonstrates that twenty-first-century-mindset leaders have been in existence for some time, as is the case with Prime Minister Lee Kwan Yew, who understood the need to prioritize the planet and people over profit. It also confirms that a nation's leader must be responsible for driving the solution and not delegate the task to anyone else. The combination of economic growth with the increase in forest / green cover and placing a priority on the water supply is clearly demonstrated as successful, as today, Singapore is considered a global hub for water and waste management, while driving its economic growth. Its twenty-first-century leaders have prioritized the planet and people over profit.

All health and social emergencies need urgent attention, and each organization must discern if it is directly or indirectly contributing to these emergencies.

4.5.5 Deforestation and Pandemics

Deforestation directly impacts the natural habitat of many species. We have wiped out more than 80 percent of primary forest cover and 60 percent of all species across the globe. Then a virus was transmitted to human beings in a wet market. Closing these wet markets and encouraging all companies to adopt a forest, conserve it, and reforest it is the way forward.

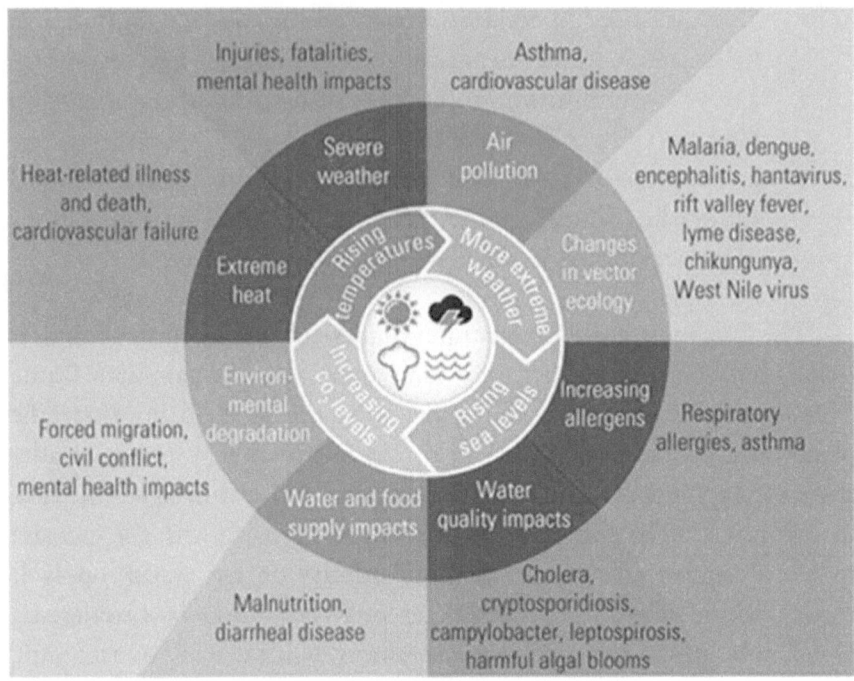

Health risks and climate change[21]

All twenty-first-century leaders will appreciate the science and understand how extreme weather, rising temperatures, increasing CO_2 levels, and rising sea levels directly affect human health. It is therefore important that twenty-first-century leaders ensure that their companies become part of the solution to all these issues and not part of problem.

4.5.6 Air Pollution and Health[22]

Every business contributes to air pollution, which kills more than seven million people each year according to WHO. All companies need to transition away from unsustainable fossil-fuel-driven logistics fleets and move to sustainable hybrid and electric fleets. How they source electricity and energy, discussed in the climate emergency module, will also make a direct impact on air pollution.

Some nations persist in using coal for their energy needs, which is most irresponsible as this immediately creates health issues. Air pollution is caused by the burning of fossil fuels and directly impacts a nation's air quality, especially in its cities.

Air pollution is a silent killer. How does a company contribute to this problem, either directly or indirectly? If your business is obtaining electricity from a grid powered by coal, will you consider making a change to a renewable source of energy? If your company's fleet of vehicles and logistics operations are powered by pollution-causing diesel or petroleum, will you consider making the change to hybrid and electric vehicles so your company may be part of the solution to climate change?

Twenty-first-century leaders walk the talk and are sensitive to the need to cease using pollution-causing fossil fuels all along the supply chain. They also set the tone by driving sustainable electric or hybrid vehicles and making it compulsory for employees across the organization to do the same.

4.5.7 Toxic Waste and Industrial Pollution

Every business needs to have policies and actions in place for eliminating toxic waste and industrial pollution, both in its operations and, more significantly, in its outsourced supply chains, wherever they may be. China and India, over the past decade, have taken a greater share of the manufacturing pie and are thus directly contributing to the crisis. This is in addition to the many oil spills that destroy pristine environments. Every fossil fuel company should be held legally accountable for the restoration of such destroyed areas based on the principle of the polluter pays.

4.5.8 Agricultural and Food Waste in Supply Chains

Even though companies worldwide have invested a significant amount of resources in agriculture, more than 40 percent of agricultural produce is lost in the distribution chain way before it gets to the end consumer, and then more than half of what reaches the consumer is also thrown away as waste because it has passed the use-by date. The world does *not* have a food scarcity problem, but it does need to fix the food waste problem by performing life cycle studies that identify the issues at each stage of the distribution cycle.

4.5.9 Single-Use Plastic Pollution, Biodegradability, and Microplastics [23] [24]

The Great Pacific Garbage Patch, which is full of used plastic products, is confirmation that we recycle only 4 percent to 5 percent of single-use plastics, with 95 percent to 96 percent of all single-use plastics ending up in the oceans or a landfill. Transitioning away from single-use plastics should be a focus of all twenty-first-century leaders to act upon.

Companies such as Coke, Pepsi, Danone, Nestlé, Mars, and Unilever use a great deal of single-use plastic, which when discarded ends up clogging up rivers and going on to become part of the Great Pacific Garbage Patch. The next time you hear of a company greenwashing about its recycling credentials, consider refusing to support that company.

Does your company use single-use plastic packaging? If so, can this be changed? For the sourcing of products, consider developing a sustainability policy that focuses on rating all suppliers in terms of their sustainability credentials.

4.5.10 The Social Emergency of Rampant Inequality

Every business can affect everything within its sphere of direct and indirect influence by investing in its employees first and encouraging them to live sustainable lifestyles. Once this is achieved, and only when it is achieved, should the business move beyond that realm to impact the greater community and society. Many companies hide behind a few CSR (corporate social responsibility) activities that are more PR and publicity initiatives than anything else.

The twenty-first-century leader is focused on directly impacting his or her organization's employees first and inspiring them to live sustainable lifestyles, as is the case with CDB Advance and the Dilmah tea company, demonstrating that any company with a twenty-first-century leader committed to contributing toward social progress can persuade its employees to live sustainable lifestyles while pursuing sustainable livelihoods.

The social emergency is as a result of unprecedented inequality. We are in a situation where 1 percent of the world's population is control more than 52 percent of the world's wealth.[25] This is why the 99% movement gained momentum. Each organization should focus on, instead of engaging in frivolous greenwash CSR activities, minimizing the disparity among its employees and the communities it operates in, thereby making a positive impact on everything within its sphere of influence. Questions to ask are as follows: What are the challenges faced by the direct employees and indirect employees when working in the various supply chains the company depends on for products and services? Can the organization encourage all employees to live sustainable lifestyles, and will it them to move to renewable energy and electric- or hybrid-powered transportation? It is significantly better to put one's house in order before reaching out to affect the outside world. It is imperative to first look inward and make an impact on one's organization's sphere of influence before getting into CSR activities, which amount to simply doing good in order to look good.

The Dilmah case study demonstrates that when a leader has a twenty-first-century mindset, he or she will give equal priority to contributing toward social progress and improving the environment by investing 15 percent of the company's profits in efforts that will positively affect the employees and the communities the company operates in.

4.5.11 Merrill J. Fernando, Founder of Dilmah Tea—Business as Means of Human Service [26] [27]

Dilmah, the largest exporter of branded Ceylon tea from Sri Lanka, founded in 1988, strives to create a better world by bringing the best from bush to cup. Despite supply chain disruptions in the past few years, Dilmah has performed well. From 2019 to 2020, revenues grew by 8 percent to US$66 million and operating profits increased by 25 percent to around US$10 million,[28] this even though in 2021, Dilmah Ceylon tea was priced 10 percent higher than the same tea from the company's direct global competitors. This is because the brand remained strongly committed to its environmental and social values and ensured high-quality natural goodness in every cup of tea.

The brand was founded by the twenty-first-century, science-led, sustainability-mindset leader Merrill J. Fernando in 1988. Merrill dedicated his life to tea and was the first producer to offer tea handpicked and packed at the point of origin, ensuring great taste and natural goodness.

At the time, Merrill's dream was for the company to give back to the community. It is for this reason the group invests a minimum 15 percent of its pretax profits in humanitarian and environmental initiatives through the MJF Charitable Foundation (to which every Dilmah tea drinker contributes) and the Dilmah Conservation and Sustainability Unit (DCSU).

The company legacy is now carried by Merrill's son, Dilhan Fernando, the CEO of Dilmah (chairman of the UN Global Compact Network, Sri Lanka). Dilhan has kept environmental and social aspects central to the brand, while leveraging on partnerships. For instance, Dilmah became a founding partner of Biodiversity Sri Lanka (BSL), an active member of the United Nations Global Compact. In terms of the environment, Dilmah has committed to the Science Based Targets initiative (SBTi) and has launched its Carbon-Negative Action Plan 2030. Social impact is important to the company, which regards employees as a valuable business resource. With a staff retention rate of 80 percent, the company has spent US$178,000 to train and develop its staff. The MJF Charitable Foundation, to date, has reached more than twelve thousand five hundred families in Sri Lanka.

Dilmah's leaders have demonstrated to the world the following twenty-first-century science-led leadership traits:

- an appreciation for mobilizing science as part of the business strategy, by adopting the Science-Based Targets initiative
- an urgent need to make an impact on the code red for business by 2025, by committing to being carbon-negative by 2030

- a triple-bottom-line mindset with the MJF Charitable Foundation / climate research center
- a willingness to embed the Twenty-First-Century Board Leadership Model in the business's board/C-suite/operations agenda, putting people and the planet before profit.

4.6 The Economic Emergency

An appropriate response to the economic emergency is to do the following:

1) **Reduce one's business's carbon footprint to be carbon-positive.**
   Twenty-first-century business leaders will be focused on monitoring the carbon footprint of their organizations as a result of their greenhouse gas / CO_2 emissions (Scope 1, 2, and 3) and on creating a strategic focused game plan to drastically reduce and minimize their organizations' contribution to the climate emergency and to the code red for business by 2025.

2) **Install a sustainable business model.**
   In order to achieve this goal, twenty-first-century business leaders must review their organizations' current business model, including every element of its strategy. A twenty-first-century leader will examine base products and services, sourcing and supply chain, branding and marketing, circularity, and target markets from the perspective of sustainability and create a sustainable business model such as that created by Patagonia.

3) **Drive sustainable innovation.**
   In order to have a sustainability-led business model that strives to reduce the organization's carbon footprint, the business must be committed to sustainable innovation and replace all its unsustainable products, services, and processes with sustainable

options. In order to do that, the business will need a pipeline of sustainable innovation.

An economic emergency does not happen in isolation but is a direct result of having ignored the climate emergency and the health and social emergency. Nineteenth-century science deniers in positions of leadership and influence have ignored the science behind the climate emergency and the health emergency and therefore have acted neither decisively nor with urgency to be part of the solution. Instead, they have gone on with business as usual. It must be added that if any business ignores its potential disruptors, it can bank on soon experiencing an economic emergency.

A twenty-first-century leader is someone who understands the need to assess the potential economic impact of each of the six modules and who takes preemptive and corrective action to minimize the economic impacts on the business model and the supply chain, which then affects the economic performance of his or her nation or business.

4.6.1 Reduce the Carbon Footprint

Any business that has a twenty-first-century leader and board will strive to minimize its carbon footprint and move aggressively to become a key part of the solution to the climate emergency, as opposed to being part of the problem. Leaders with a twenty-first-century, science-led, sustainability mindset will determine and then verify the organization's carbon footprint in terms of Scope 1, 2, and 3 emissions, which cover its supply chain, and develop a focused priority UNSDG strategy to reduce that footprint. Every twenty-first-century leader will know the carbon footprint of his or her company and will have an aggressive, urgent, and focused strategy to reduce that footprint and ensure the company or organization is not a part of the climate change problem, but is part of the solution by 2025.

The best way for any business, organization, or nation to achieve this goal is to embed the following five priority UN Sustainable Development Goals in its strategy. There is no way every business and nation can invest in advancing all seventeen UNSDGs simultaneously. The prioritized UNSDGs are focused on reducing the organization's carbon footprint, ensuring the organization's strategy leads to a positive impact on the climate emergency, and ensuring the organization is a key part of the solution to climate change.

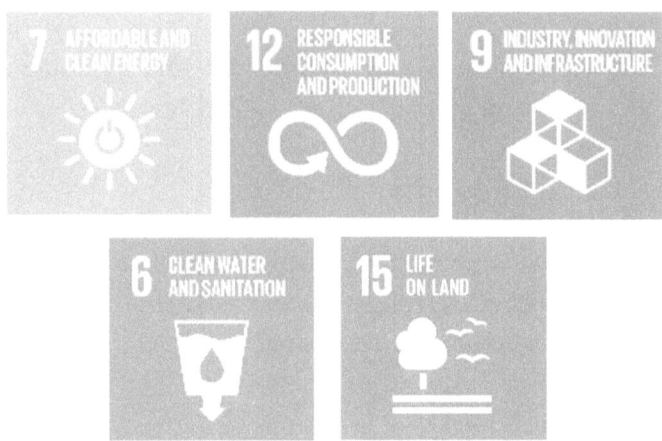

Focus on priority UNSDGs to fight the code red for business

The following UNSDGs are the ones I have prioritized for the purposes of *Twenty-First-Century Leadership to Fight the Code Red for Humanity*, including my reason for choosing each:

1) UNSDG 7

   We need to put the fire out! Global warming is caused by the burning of fossil fuels, and unless and until we transition away from using fossil fuels, no strategy we can create to halt global warming will work. Nothing has worked the past thirty-five years. Focusing every business and nation on UNSDG 7 ("Move to renewable energy and sustainable logistics/transportation") should be a priority, with each committing to a plan to end reliance on from fossil fuels by 2025 ideally or 2030 at the very latest. Every business and nation that commits to ending its reliance on fossil fuels after 2030 could be led by a nineteenth-century, science-denying, zero-sustainability-mindset nonleader who is overwhelmed by the perceived need to persist with fossil fuels and is making commitments that are at best frivolous and at worst meaningless, only to check off a box, with no intention of putting the fire out.

2) UNSDG 12

Achieving sustainable consumption by ensuring sustainable supply chains and stewardship of natural resources is the second priority.

3) UNSDG 9

Sustainable innovation and infrastructure is the third priority, with the aim of driving *only* sustainable innovation that is science led and that affects sustainability.

4) UNSDGs 6 and 15

Water and reforestation to achieve biodiversity is a joint fourth priority to ensure we halt deforestation, which has already resulted in the devastation of 70 percent to 80 percent of primary forest cover and rain forests and, with this, the loss of more than 60 percent to 70 percent of all species.

4.6.2 Install a Sustainable Business Module

In this module, we explore the business models of Patagonia, Unilever, and Tesla and come to understand the global expansion of the electric vehicle market.

As a guiding principle, twenty-first-century leaders who are science led with a sustainability mindset will deliver above-average medium-to-long-term economic performance by adopting twenty-first-century business models that put the planet and people over profit and, by so doing, will deliver excellence in terms of sustainable innovation and profitability.

Nineteenth-century science-denying leaders will deliver short-term profits, but in so doing they will significantly compromise the planet for future generations. The benchmark companies are, or were, led by exemplary twenty-first-century leaders with a strategic mindset. Yvon Chouinard, founder of Patagonia; former Unilever chairman Paul Polman; and visionary leader of Tesla, Space X, and Solar City Elon Musk are the pacesetters. In the automotive industry, we see that by 2022 Tesla completely overwhelmed the market capitalization of all its industry peers.

This clearly demonstrates that a twenty-first-century leader can, by doing the right thing, also do well. The cases showcased here all demonstrate the commitment to the four traits that is made by twenty-first-century leaders who are led by science and have a sustainability mindset. These people were the first who were ready to take a bold step to go against the norm of creating a business model that did nothing more than ensure profits to shareholders at any cost.

The conviction that each of these people had to prioritize the planet and people over profit is what sets all of them apart. The other key distinguishing factor is that these leaders were all deeply committed to being a part of the solution to the climate emergency and code red for business, not just disengaged onlookers.

The twenty-first-century, science-led, sustainability-mindset leader Yvon Chouinard set the direction Patagonia would take and the values it would operate by at the very outset, in 1973, by ensuring that every single decision Patagonia made would prioritize the planet and people over profit, as demonstrated below.

4.6.3 Yvon Chouinard, Founder of Patagonia—Leading Circularity in the Clothing Industry[29] [30] [31] [32] [33] [34] [35] [36]

Patagonia Inc. is an outdoor clothing company founded in 1973 by Yvon Chouinard and based in Ventura, California. The company operates more than seventy stores worldwide (while encouraging online retailing) and has offices in the United States, the Netherlands, Japan, South Korea, Australia, Chile, and Argentina. Patagonia has a unique sustainable business model based on the premise of "Buy less, demand more." The brand was voted as the most reputable brand in the United States according to Axios Harris, which polled a nationally representative sample of nearly forty-three thousand Americans.

The environmental heritage of Patagonia began with the founder, Yvon Chouinard. As a twenty-first-century science-led leader and environmentalist with a sustainability mindset, he reasoned in his catalogue essays in 1974, "No longer can we assume the earth's resources are limitless. ... Mountains are finite, and despite their massive appearance, they are fragile." He added meaning to his business (at the time by making and selling reusable rock-climbing equipment to his friends and fellow climbers) by aiming to make Patagonia's clothing last a lifetime by manufacturing it with high-quality raw materials and ensuring it was repairable. As a core principle, Chouinard chose to give back. At the start of his business, he tithed 10 percent of his profits to grassroots environmental organizations, later expanding this to 1 percent of all sales, calling it an "earth tax."

The legacy is now carried on by current CEO Ryan Gellert, whose main mission is to save our home planet. Overall, business is smooth with annual sales in 2021 of close to one billion US dollars. The current climate figures for the company were also impressive as of 2020, the company operating on 100 percent renewable electricity in the United States and 76 percent globally (aggressively moving toward 100 percent). In terms of sourcing in 2021, 100 percent of the virgin cotton was grown organically. When it comes to recyclability, 87 percent of the fabrics are made with recycled materials. The company also has free guides for repairing Patagonia products, having repaired 50,295 articles of clothing in 2017. The actions the company takes do not stop there. Gellert's vision is for the following things to happen:

- ✓ By 2025, Patagonia will be 100 percent carbon-neutral across their entire supply chain.
- ✓ By 2025, all apparel products will be made from 100 percent recycled, reclaimed, or renewable resources.

✓ By 2025, packaging will be 100 percent reusable, home compostable, renewable, or easily recyclable (making Patagonia a zero-waste-to-landfill company).

Although the fashion industry is expected to miss the +1.5°C target by 50 percent in 2030, Patagonia has demonstrated that by having courageous leadership, promoting mindful spending, and offering high-quality products, a company can make a positive impact on people and the planet while being rewarded with high profits.

Yvon Chouinard has demonstrated to the world the following twenty-first-century science-led leadership traits:

- an appreciation for mobilizing science as part of the business strategy, by leading his company to become 100 percent renewable energy driven in the United States and 76 percent globally
- an urgent need to make an impact on the code red for business by 2025, with all climate targets to be achieved by 2025
- a triple-bottom-line mindset, prioritizing the planet over profit, donating 10 percent of Patagonia's profits to grassroots environmental organizations and 1 percent for the planet
- a willingness to embed the Twenty-First-Century Board Leadership Model in the business's board/C-suite/operations agenda, with results achieved since 1973.

4.6.4 Paul Polman, Former CEO of Unilever—Unilever Sustainable Living Plan [37]

Unilever is a global FMCG company with more than four hundred brand names in more than one hundred ninety countries, with a hundred-year legacy. The company became a frontrunner in sustainability under the twenty-first-century, sustainability-mindset, science-led leader Paul Polman. The now former CEO came onto the scene in 2009, and one of his first actions was to abolish quarterly reports (shifting to a long-term focus to help tackle the code red for humanity and business) and earnings guidance. Three years later, in 2012, he created the Unilever Sustainable

Living Plan (USLP), a decade-long scheme to double the company's revenue while considering its role in environmental stewardship and doing business the right way. Unilever embraced sustainability and embedded it in its corporate strategy for the first time under Paul Polman in 2009. He championed the cause, despite opposition from nineteenth-century board and C-suite leaders, until his departure in 2019. Among the successes of the USLP after one decade had passed in 2021 are the following items:

✓ more than €1.2 billion (US$1.37 billion) in costs avoided
✓ helped 1.3 billion people improve health and hygiene
✓ became 100 percent renewable in terms of electricity used for manufacturing
✓ saw a 65 percent reduction in CO_2 from energy used in manufacturing
✓ achieved global gender parity in management at 51 percent
✓ saw water use go down to 49 percent per ton of production
✓ 67 percent of agricultural raw materials sustainably sourced
✓ zero waste to landfill for all factories.

Paul Polman continuously pushed business leaders to care about the world's problems. His twenty-first-century mindset promoted the fact that net-positive leaders who are restorative, reparative, and regenerative to nature are also societal leaders. Paul leveraged the expertise of several organizations (i.e., the multistakeholder model) to meet targets and expand his business sustainably. In this, he has demonstrated to the world the following twenty-first-century science-led leadership traits:

• an appreciation for mobilizing science as part of the business strategy, by executing the USLP in 2012
• an urgent need to make an impact on the code red for business by 2025, by achieving 100 percent renewable energy used in manufacturing by 2021
• a triple-bottom-line mindset, prioritizing the planet over profit
• a willingness to embed the Twenty-First-Century Board Leadership Model in the business's board/C-suite/operations agenda. See the USLP results achieved from 2009 to 2021.

4.6.5 The Twenty-First-Century Leadership of Elon Musk of Tesla, Space X, and Solar City [38] [39] [40] [41]

The twenty-first-century visionary leadership of the science-led, sustainability-mindset Elon Musk is clearly a benchmark case that demonstrates how making a strategic commitment to prioritize the planet and people over profit can result in exceptional business returns in the medium term.

Elon Musk, the CEO of Tesla, is a visionary in terms of leveraging science-led leadership in sustainable automation and space exploration. Tesla, his primary concern, is focused on accelerating the world's transition to sustainable transportation and energy.

Road transportation accounts for 10 percent of global emissions. According to the International Energy Agency (IEA), in order to reach net zero by 2035 (as forecasted, by 2030 there will be 120 million EVs), no more vehicles powered by fossil fuels should be sold, and 50 percent of trucks must be electric. In 2020, the global fleet of Tesla vehicles and solar panels enabled Tesla's customers to avoid emitting five million metric tons of CO_2.

Tesla, in addition to being a global electric vehicle company, is also a clean energy company, manufacturing not only electric cars but also battery energy storage units from home scale to grid scale, solar panels, and solar roof tiles. Headquartered in the United States, it was founded in 2003 with five factories (in Nevada, New York, Shanghai, Berlin, and Texas). In October 2021, with Hertz's commitment to purchase one hundred thousand Teslas, the company became one of five companies valued at more than one trillion dollars (the second-fastest company ever to reach that mark, only eleven years after its public debut in 2010).

As of January 2022, Elon Musk became the richest person in the world with a net worth of US$269 billion. Musk's main aim is to revolutionize transportation both on earth, through electric car maker Tesla, and in space, via rocket producer SpaceX. His strategy is to create a complete energy (Powerwall, Powerpack, and Solar Roof) and transportation ecosystem.

Tesla seeks to achieve this goal by spending large amounts on R & D and software development efforts and by developing advanced manufacturing capabilities. For instance, Tesla spends an average of $2,984 on R & D per car sold and $0 on advertising.

Tesla also manufactures a unique set of energy solutions, enabling homeowners, businesses, and utilities to manage renewable energy generation, storage and consumption. Because SpaceX and Tesla are both headed by Musk, Tesla can leverage space technology, autonomy, robots, and artificial intelligence at scale for use in its own products.

Elon Musk has demonstrated to the world the following twenty-first-century science-led leadership traits:

- an appreciation for mobilizing science as part of the business strategy, as evidenced by his disruptive, science-led leadership in futuristic space technology, automation, and renewable energy solutions for the long term
- an urgent need to make an impact on the code red for business by 2025, as shown by the launch of Tesla and Solar City
- a triple-bottom-line mindset, prioritizing the planet over profit in all decisions
- a willingness to embed the Twenty-First-Century Board Leadership Model in the business's board/C-suite/operations agenda. See his results achieved from 2003 to 2022.

Every one of the leaders featured in the Patagonia, Unilever, and Tesla cases live by and practice all four traits that differentiate them as twenty-first-century leaders. All of them are science led. Elon Musk has a PhD in physics and has ensured every one of his business initiatives is going to be a solution to the climate emergency. Tesla has set the pace for accelerating the position of electric vehicles in the global marketplace with six factories, each with a capacity for five hundred thousand vehicles. Space X is setting a new paradigm in reusable rockets, and Solar City is accelerating solar plus battery (Powerwall) solutions. All of these ventures prioritize the planet and people over profits, confirming that Musk is a sustainability-mindset leaders who walks the talk of governance. The key decisions that he and the others mentioned have made, along with the results after implementing those decisions, confirm that each has an attitude of zero compromise, ensuring that all decisions conform to the twenty-first-century principle of governance. All of them have demonstrated an urgent need to make an impact on the climate emergency and on the code red for humanity and business by launching aggressive programs to be a part of the solution, and each has provided the leadership required to align the board, the C-suite, and the operations team to drive the twenty-first-century leadership agenda. Patagonia has set the global benchmark for sustainable apparel, with repairability as a radical component that sets the company apart and challenges every apparel company to move toward long-lasting clothing, not the type that people wear a few times and then throw away. Unilever set a new paradigm by refusing to report quarterly results and moving toward annual and strategic results, a key paradigm for sustainable business. Its progress in the past ten years, up to the time Paul Polman left the organization in 2019, has been phenomenal. However, it is clear that Unilever's board of directors were not as enlightened as Polman was and had no twenty-first-century mindset.

Tesla has set a blistering pace for the manufacture of sustainable automobiles. The single-charge range of more than four hundred kilometers is already the industry benchmark, along with the vehicles' modern design and space technology.

These leaders have demonstrated that one can indeed be a twenty-first-century leader and beat the competition on a global scale.

4.7 The Disruptors

Twenty-first-century leaders will certainly place first priority on making a direct impact on the three emergencies. They will do so to ensure that their business or organization is no longer directly contributing to the code red for humanity and business and is no longer the cause of the problem in any way. At the very least, every business should minimize the negative effects of its operations on the planet and on people and significantly increase the positive impacts in order to deliver net-positive value to the planet.

The next key challenge for twenty-first-century, science-led, sustainability-mindset leaders is to minimize the impact of the three global disruptors. A disruptor is any global trend that has the ability to disrupt business as usual unless strategically anticipated and acted upon proactively.

While the three emergencies require leaders to act with urgency, every disruptor requires close study and analysis in order to understand its potential to disrupt the organization and in what ways it may do so. The three key disruptors all have the ability to disrupt the business significantly. Early anticipation of these disruptions, which present strategic opportunities as well as risks, could allow a leader to avert them or, at the very least, minimize them. Here I am speaking of technology disruption, geopolitical disruption, and governance disruption, all of which can disrupt a business and make it irrelevant in the twenty-first century.

Technology can disrupt a business if the business fails to anticipate more-efficient technological advancements that will give certain competitors a significant advantage over all the others, thereby threatening to make the competition irrelevant. In the past decade, we have seen many such disruptive technologies. The mobilization of AI, blockchain, robotics, 3D printing, precision agriculture, advanced solar panels plus batteries, smart grids, and electric sustainable transport has directly disrupted many business sectors all over the world.

Electric vehicles, agriculture, tourism, and hospitality are a few industries where a new set of global companies and nations are leading the sector. Here I am thinking of Uber, Airbnb, Tesla, and Google, and even precision agriculture led by Holland, which today is the second-largest exporter of agriculture products in the world.

Geopolitical disruption can have devastating effects on a business. During the Trump administration, the United States commenced a trade war with China. Many US and Chinese multinational corporations (MNCs) were affected, for example, Huawei and TikTok. In March 2022, we have seen the mass exodus of many US MNCs from Russia—leaving as a sign of protest against the Russian invasion of

Ukraine. China's One Belt One Road initiative is currently responsible for giving China a strategic advantage over the USA and the European Union—and at least one hundred thirty-eight nations. The emerging alliance between China and Russia could have far-reaching consequences for all MNCs and transnational organizations.

To disrupt governance is an intentional decision made by nineteenth-century leaders who deny science and seek profit at any cost, even prioritizing profit over the planet and people. This mindset is clearly demonstrated in the fossil fuel industry and the banking sector.

In both instances—geopolitical disruption and governance disruption—we see that companies such as Saudi Aramco, Gazprom, BP, Shell, Total, and Chevron, and banks such as JPMorgan, Citigroup, Bank of America, HSBC, Barclays, and Standard Chartered, have all compromised the planet by putting profits ahead of the environment. The Banking on Climate Chaos Report 2021 confirms this compromise. We see the EIB (European Investment Bank) and Santander Bank as exceptions to the rule, with twenty-first-century science-led board members and C-suite occupants.

Twenty-first-century leaders need a strategic approach to manage every single disruption. They need to be science led to embrace the future of technology, and they need to be brought up to speed with every geopolitical trend that will affect every multinational and transnational business. They need to have a sustainability mindset to understand the new paradigm for governance, where every decision must place priority on the planet and people over profit.

The disruptors need to be viewed as making an impact in the medium to long term, versus the short term, which is where the emergencies make their impacts. The three disruptors need to be strategically assessed in terms of their potential to impact to the business model or supply chain. It must be accepted that no board, C-suite, or leadership team has within itself the knowledge, expertise, and skills to address these three disruptors competently.

Board advisers who are well versed in each of the disruptors will need to make an independent assessment of the potential opportunities and risks to the business/industry. The key is to anticipate the disruption by reading the trends and to build a well-informed twenty-first-century strategy that, where possible, takes advantage of potential opportunities and minimizes the risks.

4.7.1 The Technology Disruption Module

Technology disruption soon will be a reality that every twenty-first-century leader will be sensitive to and knowledgeable about. All twenty-first-century technologies that have the potential to disrupt a business model and embrace it are first movers.

Nineteenth-century leaders will either be ignorant of or fail to pay attention to the major technological advances and will persist with the existing technology, which ultimately will cause delays. No person should be in a position of influence, whether local, national, or global, if he or she is clearly a nineteenth-century leader.

Our focus in terms of the technology disruption arena are those technological advances that will make a positive impact on the climate emergency and the health and social emergency in terms of carbon emissions. Therefore, the key technologies needed by twenty-first-century leaders who are committed to fighting the code red for business are as follows:

- twenty-first-century solar power grids
- AI mobilization
- nanotechnology, water, and graphene water desalination
- precision agriculture
- blockchain
- 3D printing, robotics, and nano–bio/

The foregoing is only a select list. The key point is to be aware of any technology that has the ability to disrupt the business model. Twenty-first-century leaders will be early adopters and not laggards when it comes to embracing new technology before it can disrupt business models and supply chains.

This is an area where it is advisable to contract an independent expert in the industry who can advise the organization as to which of these technologies may make a direct or indirect impact on the business and which, therefore, need to be invested in.

The twenty-first-century leader is apprised of and well versed in technologies that are on the horizon and makes calculated decisions as to which technologies to mobilize in his or her business.

Certain technologies will directly help businesses and nations be part of the solution to the code red for humanity and business. Among these are technologies that can be mobilized to put an end to global warming are technologies that are supported by UNSDG 7 in terms of embracing renewable energy versus persisting with global-warming-causing fossil fuels.

4.7.2 Power Grids of Tomorrow (Fluence Energy) that Leverage AI

The focused solution to the fossil-fuel-driven energy problem is to take the first step to reduce global warming and put an end to the climate emergency. As long as businesses and nations persist with using fossil fuels, we will never find a solution to the climate emergency or to the resultant extreme weather that will devastate the planet.

The first priority of a twenty-first-century business leader must be to ensure the business is not directly or indirectly contributing to the climate emergency. This is done by transitioning energy and transport from pollution-causing fossil fuels to renewable sources.

The world is moving toward twenty-first-century grids that are renewable, combining solar and wind with battery storage and AI-driven solutions. Fluence energy is setting the pace for renewable-energy-driven, battery-storage-enabled, and AI-driven solutions that are both efficient and effective. The world is also moving rapidly toward making sustainable energy solutions that are cost-effective and are effective options to pollution-causing fossil fuels.

Improvement to solar panel efficiency is another groundbreaking development. The new perovskite solar panels show a significant improvement in terms of efficiency, from 16 percent to 19 percent efficient to more than 30 percent efficient [42].

4.7.3 AI Mobilization

Google, led by twenty-first-century CEO Sundar Pichai, is a company that has made a focused commitment to leverage AI to drive the business and give it a significant advantage as shared in the case study that follows:

The Google Case and Its 24/7 Commitment to Making the Energy Transition

4.7.4 Sundar Pichai, CEO of Alphabet—Renewable Energy Transition of Google [43] [44] [45]

Google—24/7 Carbon-Free Energy by 2030

The main mission of Google is to organize the world's information and make it universally accessible and useful to everyone. The tech giant, which was founded in 1998, and is now among the four largest companies in terms of market cap. The company boasted $76 billion in profit for 2021, while aggressively pursuing climate targets.

Under the twenty-first-century, science-led, sustainability-mindset leader Sundar Pichai, who took over as CEO of Google in 2015, the company is focused on environmental sustainability as a core value to reduce its global carbon footprint. Google became carbon-neutral in 2007 (operations still create carbon, but this is compensated for by offsetting).

Sundar Pichai has made a commitment that by 2030, Google will by 24/7 carbon-free (Google will operate using clean energy twenty-four hours a day). In addition to making the commitment to be the first major company to become carbon-free by 2030, he has pledged to achieve the following:

✓ Eliminate all Google's legacy carbon emissions since 1998.
✓ Deploy $5.75 billion in sustainability bonds in Europe and globally.

✓ Enable five gigawatts of new clean energy in manufacturing regions by 2030.
✓ Launch a €10 million Climate Impact Challenge to accelerate Europe's green recovery.
✓ Help more than five hundred cities to reduce their carbon emissions by one gigaton annually by 2030.
✓ Remove more carbon from the atmosphere with reforestation and better tree data.
✓ Create breakthrough AI to dramatically increase building efficiency.
✓ Offer one billion people new ways to take actions to reduce their environmental footprint by 2022.
✓ Create pathways and opportunities for green jobs in Europe.

Considering the foregoing, Sundar Pichai has demonstrated to the world the following twenty-first-century science-led leadership traits:

- An appreciation for mobilizing science as part of the business strategy. In 2020, Google again matched 100 percent of its global electricity use with purchases of renewable energy and commitment to AI-driven solutions.
- An urgent need to make an impact on the code red for business by 2025, by making the commitment that Google will be 24/7 carbon-free by 2030.
- A triple-bottom-line mindset, prioritizing the planet over profit.
- A willingness to embed the Twenty-First-Century Board Leadership Model in the business's board/C-suite/operations agenda—Results achieved from 2016 to date.

4.7.5 Precision Agriculture and the University of Wageningen Case

As global warming and extreme weather incidents devastate the planet, we will see an immense impact in terms of freshwater scarcity across the globe with the melting of the Arctic, Antarctic, and Himalayan ice caps and as extreme weather incidents such as droughts, floods, and extreme heat destroy mass agriculture.

We will have an abundance of salt water and little or no fresh water as this happens. As more than 70 percent of fresh water is being used in agriculture, this will directly affect the future of mass agriculture as we know it.

The rising global temperatures will significantly increase the number of extreme weather incidents, which are already devastating economic assets and agriculture. Extreme weather incidents have increased to more than four hundred per annum and are threatening to destroy mass agriculture with unbearably hot temperatures, droughts, and floods, giving rise to the possibility of global food scarcity and food insecurity.

It is with this background that we look at the twenty-first-century leadership of Wageningen University & Research, which has already positioned itself as the Silicon Valley of precision agriculture. Precision agriculture—a climate-controlled, AI-driven, water-efficient (90 percent less water) solution to high-technology-driven agriculture—occurs mostly in high-rise greenhouses, which can be more resilient to extreme weather than mass agriculture. The progress Holland has made with precision agriculture confirms that this advanced technology, which was mobilized in this sector because of Wageningen University's ecosystem for advanced agriculture, has made the extremely small landmass of Holland significantly more productive and more resourceful than most of the nations where a significant amount of land is allocated to agriculture, including the USA, Brazil, Canada, India, China, and Ukraine. Vertical agriculture coupled with AI-driven inputs and a 90 percent savings of freshwater resources, along with the mobilization of hydroponics and aquaponics, has made the nation the most advanced in terms of precision agriculture mobilization.

4.7.6 Wageningen University & Research—the Silicon Valley of Precision Agriculture[46] [47] [48]

The Netherlands has become a hub for precision agriculture over the last decade. The Dutch are the second-largest producers and exporters of agricultural products in the world behind the USA and are also the largest exporters of potatoes and onions. The thrust behind this transformation is several educational bodies/universities that have nurtured twenty-first-century leaders in the fields of agricultural technology and experimental farming. Wageningen University & Research (WUR), thought to be the world's leading agricultural research institution, is located fifty miles southeast of Amsterdam.

The university was named the most sustainable institution of higher education in Netherlands by *Sustainability* magazine in 2021. WUR's vision is to disseminate high-quality, up-to-date scientific and academic knowledge to help its more than thirteen thousand five hundred students (with more than two thousand PhD candidates) become "T-shaped" specialist professionals. In the domain of food production, nutrition, and environmental issues, WUR research is clustered in the areas of food and agrotechnology, environmental research, bio-based research, livestock research, animal sciences, and food safety research.

The university is action- and partnership-oriented (its main stakeholders are end users, suppliers, chain parties, and knowledge institutes) with more than one thousand projects in more than one hundred forty countries, with formal pacts with governments and universities. These partnerships are mainly in the domain of precision agriculture. Smart farming / precision agriculture means that plants get precisely the treatment they need (which allows production to be optimized), determined with great accuracy (actions for each square meter of crop) thanks to the latest technology (GPS, sensor technology, ICT, robotics). The twenty-first-century college dean Dr. J. E. (Ernst) van den Ende mentions that he wants WUR research to be "science-driven in tandem with the market-driven," which is the only way it "can meet the challenge that lies ahead."

WUR has demonstrated to the world the following twenty-first-century science-led leadership traits:

- an appreciation for mobilizing science as part of the business strategy, by being a science-led organization
- an urgent need to make an impact on the code red for business by 2025, by developing precision agriculture to replace mass agriculture affected by extreme weather
- a triple-bottom-line mindset, prioritizing the planet and people over profit with the commercialization of precision agriculture

- a willingness to embed the Twenty-First-Century Board Leadership Model in the business's board/C-suite/operations agenda, as evidenced by WUR's being the world's second-largest exporter of agricultural goods.

4.7.7 Graphene Water Desalination [49]

Global fresh water, which makes up just 3 percent of all the water on the earth, is dwindling because of global warming and the climate emergency. The unprecedented rise in global temperatures, which for the past twenty-one years have been the hottest ever recorded in history, has accelerated the melting of the polar ice caps. The idea that humanity will face a global freshwater scarcity is no longer a simple prediction, but a reality already faced by most of the Southern Hemisphere nations. One silver lining in terms of science-led solutions is the work being done at Manchester University. To face the reality of global freshwater scarcity, Andre Geim and Konstantin Novoselov of Manchester University have offered a glimmer of hope by presenting the world with a more cost-effective and advanced way to desalinate water, which is going to be a game changer for humanity. Today, most nations desalinate seawater at a significant cost to the environment as fossil fuels are required in the process. The invention of graphene, a twenty-first-century material made from graphite, could be the thing that makes the difference.

In this module, which is the most dynamic, we have covered a few topics. We now know that the twenty-first-century leader is one who is relevant and up to date in terms of all present and future technologies irrespective of their direct impact on the industry currently engaged in. This is the attitude of the future-oriented twenty-first-century leader and strategist, who understands the need to be a first mover with technology, as opposed to being a laggard.

4.8 Geopolitical Disruption Module

Twenty-first-century leaders understand that geopolitics must not be left to politicians and national leaders. Every business that has a multinational, transnational, or export footprint will mobilize board advisers to constantly analyze global trends and give expert advice on how to address both the strategic risks and opportunities.

The Twenty-First-Century Board Leadership Model includes a geopolitical module, which is mostly absent in the curriculums of most MBA and governance programs. Nineteenth-century leaders are mostly insensitive and incapable of responding to geopolitical disruption and the subsequent opportunities. Every geopolitical challenge has the potential to significantly disrupt the business model and, at times, to present an unforeseen opportunity. Monitoring every key global geopolitical maneuver that may make an impact on the business becomes an essential element of a twenty-first-century strategy.

The reason geopolitical disruption is a key module in the Twenty-First-Century Board Leadership Model is because, unlike at any other time in history, today business after business is being completely upstaged and broadsided by the geopolitical maneuvers of many nations that have haphazardly created policies and barriers to entry, which has devastated the business that is generated within their borders. If the business is led by a twenty-first-century leader with a sustainability mindset who understands global geopolitical trends, then such a person will preemptively respond to minimize the impact or seize the opportunity.

Unpredictable scenarios that are played out from time to time will be the new norm. But in most instances, a board advisory expert on geopolitics will be able to predict the scenarios that will be played out and recommend concrete actions to minimize the impact or take advantage of the trend.

Today, we see that many geopolitical strategies are being played out by the key actors China, Russia, the USA, and Europe.

4.8.1 Russia's Invasion of Ukraine in February 2022

Despite zero provocation and the condemnation of the UN and the global community, Russia has caused massive destruction in Ukraine, destabilizing the nation and setting off a flood of refugees crossing to border nations in a bid to escape Russian aggression and military attacks. Unprecedented sanctions have been imposed on Russia, and many a

Western multinational company has exited Russia in protest of its invasion. Russia's foreign reserves have been frozen, making its US$750 billion no longer accessible. The MNCs who had earned a significant amount of money from their operations in Russia have had to sacrifice sales and profits. This will increasingly be the norm when global energy prices rise to unprecedented levels, crossing US$120 to US$140 a barrel, along with food prices rising by between 30 percent and 60 percent by commodity.

Could the crisis have been anticipated? Could it have been prevented geopolitically?

The fact that Europe depends on Russia for 40 percent of its natural gas requirements has hamstrung the aggressive response of ceasing to use Russian fossil fuels. This has highlighted the urgent need to transition away from using fossil fuels and embrace renewable and nuclear energy as an alternative.

Understanding the idiosyncrasies of nineteenth-century, science-denying, sustainability-ignoring, truth-denying, humanity-devastating nonleaders who are in positions of influence will be a key ability of twenty-first-century leaders who have an understanding of geopolitics. Monitoring their comments, actions, and intentions will prepare companies to be proactive, versus reactive to these nineteenth-century nonleaders.

4.8.2 The European Union's Green Deal [50] [51]

Europe's commitment to sustainability has been jolted by the Russian invasion of Ukraine, forcing the European Union to end its dependence on Russian oil and gas by 2027. It is clear that the EU is the only region focused on the 2015 Paris Agreement, with many twenty-first-century leaders. The commitments made by the EU to build sustainable supply chains to achieve the goal of a zero-pollution Europe are ambitious and need the support and commitment of twenty-first-century science-led leaders with sustainability mindset leaders if they are to be achieved. Pursuing a transition to a circular economy, from farm to fork, will require significant investments in supply chain development and sustainability thinking. Significant investments of money to help achieve the transition to renewable green energy sources and sustainable transport will be the new focus.

The USA no longer has the global influence, economic power, or defense capability to be the key influencer of the global geopolitic. The Trump regime refocused the nation inward, denied the climate emergency, and initiated a trade war with China. The Biden regime, which has reversed many of these actions since taking over in January 2021, is much more sensitive to the United States' global role; has resumed support of the Paris Agreement to make a positive impact on the climate emergency; and has changed the insular foreign policies to be more engaged.

China is increasing its global influence and access to resources with the One Belt One Road strategy, which already has 138 nations actively involved. Its focus on reclaiming Taiwan has many strategic geopolitical ramifications. China's economic and geopolitical influence is on the rise, and there is evidence that the nation is pursuing a new world order with Russia. The new China-Russia alliance seems to be focused on each of these nations advancing their strategies despite global objection. The China-Russia alliance, which will most likely be joined by Brazil, India, and Saudi Arabia, will delay the urgently needed transition to renewable energy and extend the use of fossil fuels.

4.8.3 China—One Belt One Road Initiative [52] [53] [54]

Cementing China Geopolitical Influence through a Global Silk Route

China's Belt and Road Initiative (BRI), formerly known as the One Belt One Road initiative (OBOR), is a twenty-first-century inland (belt) and maritime (road) trade network that sought to connect China to East Asia and Europe by linking 144 countries by 2021.

The project, which was launched by President Xi Jinping in 2013, is set to be one of the most ambitious infrastructure projects (investments, modernizing old infrastructure, building new infrastructure that consists of around twenty-six hundred projects in more than a hundred countries [mostly underdeveloped and developing nations]) in the world, touching around 60 percent of the world's population (the World Bank has estimated that the project has taken seven million people out of poverty). It is estimated that by 2027 the total investment could reach up to US$1.3 trillion.

For China, the BRI has drawn eight hundred forty-seven thousand Chinese nationals to work abroad in more than sixteen thousand companies. The project will expand China's economic and political influence, the former by way of obtaining raw materials needed for China's value-added goods. At COP26 in Glasgow in 2021, China committed to stop building any new coal-fired power plants. President Xi also said China will support developing countries in their move toward green and low-carbon energy by accelerating sustainable finance through its Asia Infrastructure Development Bank (AIDB).

China has demonstrated to the world the following twenty-first-century science-led leadership traits:

- an appreciation for mobilizing science as part of the business strategy, by being the world leader in nanotechnology, AI mobilization, electric rail, and solar and wind energy
- an urgent need to make an impact on the code red for business by 2025, by building infrastructure to accelerate the shift to a low-carbon economy
- a triple-bottom-line mindset, ending any coal power project outside China in BRI countries
- a willingness to embed the Twenty-First-Century Board Leadership Model in the business's board/C-suite/operations agenda, by impacting the economic growth of 144 nations through infrastructure.

Every twenty-first-century business strategy and leader needs to account for or be sensitive to every one of the foregoing moves and its likely impact on access to global markets, possible blacklisting because of the country of origin, access to global resources and financing, access to supplies, and all global supply chains, all of which need to be factored in to the corporate strategy and business plan. The twenty-first-century strategist needs to be fully conversant in how each of these key factors that impact business performance is factored into every strategy.

In addition to the potential disruption as a result of global geopolitics are two other key issues that have far-reaching geopolitical ramifications: global debt and access to fresh water.

- **Global debt** gives lending nations a strategic advantage and influence over nations that depend on them. The point must be made that while global GDP in 2021 was estimated to be US$94 trillion, today global debt has ballooned to nearly 300 percent of global GDP, nearing US$300 trillion. This has made a major impact on global consumption of resources and will contribute to accelerated consumption of limited finite resources, which will create an issue for future generations in terms of sustainable development. Global debt was estimated to be US$281 trillion as of 2020 and was set to rise again in 2021.

- **Global freshwater scarcity** is another key issue that immediately affects geopolitics. Nations that have depleted freshwater resources because of the climate emergency and global warming will be vulnerable. A majority of nations in the Southern Hemisphere are expected to be impacted by both economic scarcity and physical water scarcity by 2025. This fact immediately makes nations rich in freshwater resources and forest cover, which enables the rain cycle, to have disproportionate influence over those that don't have access to this life-giving resource.

Water Scarcity and Climate/Environmental Refugees

According to the World Bank's Groundswell report, by 2050 we could see an unprecedented increase in the number of environmental refugees with sub-Saharan Africa estimated to have eighty-six million refugees; South Asia, forty million refugees; and Latin America, seventeen million refugees.

Both global debt and water scarcity will be two key issues twenty-first-century leaders must focus on to understand their geopolitical implications. Debt gives richer nations control and influence over poorer nations, and water scarcity will prompt an unprecedented climate refugee migration from the Southern Hemisphere to the Northern Hemisphere. These two factors need to be addressed by every corporate strategy made by a business that has a global footprint.

4.9 Governance Disruption Module

The twenty-first-century leader will be a champion of the new paradigm for governance, where every decision considered will be evaluated in detail before *any* final determination is made, taking the solution with the least negative impact. Training twenty-first-century leaders to evaluate all options and decide on the strategy or option that minimizes the negative impact to the planet and its people, instead of simply rushing to make the decision, will be a key differentiating factor.

We live in a world led by nineteenth-century science-denying leaders who pursue profits at any cost. The global fossil fuel industry, the banking sector, the single-use plastics industry, and the weapons industry are a few examples of planet-compromising sectors that are run by nineteenth-century boards, CEOs, and C-suites leaders. The future paradigm is one where every decision made will prioritize the planet and people over profit.

Every profit and investment decision needs to be analyzed both financially and in terms of environmental impact assessments (EIA) and the negative impact it will make on the planet.

Any action that threatens to make a negative impact on the planet needs to be avoided at all costs. Decisions made in the name of development and growth must be evaluated by considering various scenarios. AI can be rapidly mobilized to evaluate these various scenarios, and the decision must be based on whatever will make the least negative impact on the planet.

Governance Compromises of the Century—Subsidizing Fossil Fuels (US$5.9 Trillion per Annum)

1) **Compromise 1—persisting with fossil fuels**
   Nineteenth-century leaders of nations and businesses have significantly compromised the planet by persisting with fossil fuels for the past thirty-five years, after the Bruntland report clearly highlighted the need to cease using these pollution-causing forms of energy.

2) **Compromise 2—subsidizing fossil fuels**
   Nineteenth-century leaders have denied science and compromised the planet further by investing in actions that promise to accelerate the devastation of the planet. They have invested up to US$5.9 trillion per annum to subsidize the cause of the climate emergency and the code red for humanity and business.

   This confirms the hypothesis that most leaders are ignorant of the science and simply put profits before the planet. Simply making fossil fuels more accessible and cheaper in order to facilitate lower supply chain costs are nineteenth-century science-denying decisions that have contributed to the code red for humanity and business.

   Twenty-first-century leaders prioritize the planet and people over profit in every decision. They will address the elephant in the room and correct the science denial of nineteenth-century science nonleaders by doing the following things:

   1) Transitioning away from pollution-causing fossil fuels by 2025–2030.
   2) Putting an end to the subsidizing of fossil fuels by 2025.

4.9.1 Banking on Climate Chaos Report 2021

Nineteenth-century, science-denying, profit-at-any-cost leaders in the banking industry have compromised the planet as clearly highlighted in the Banking on Climate Chaos Report. The devastation of the planet has been financed despite commitments made by the very banks that signed the 2015 December Paris Agreement and promised to stop doing so, clearly confirming that these are nineteenth-century science deniers, investing more than three trillion US dollars between 2015 and 2022.

One of the key characteristics of a twenty-first-century leader is a strategic focus when making decisions. The twenty-first-century board is like-minded, supporting strategic decisions versus short-term quick fixes. Governance disruption occurs when nineteenth-century leaders focus on short-term results, denying science and, wanting to win in the short term, sacrificing the future at any cost.

Governance compromises are rampant, with COP26 confirming that many nations with a vested interest in fossil fuels were ready to deny the scientific evidence that continuing to do so would contribute to the code red for business and dig the grave for humanity. Crossing the global temperature line of +1.5°C to +2.7°C seems inevitable between 2025 and 2030.

Nations in the fossil fuel industry or dependent on fossil fuels for energy were ready to let the code red for humanity and business unfold, leading to catastrophic devastation of the global economy, with a twenty-foot rise in the sea level, to the tune of US$54 trillion to US$67 trillion by 2030 because of extreme weather incidents and the melting of the polar ice caps.

The nine nations complicit in digging the grave for humanity are Saudi Arabia, China, India, Russia, Australia, Brazil, Indonesia, Mexico, and Turkey.

Banking on Climate Chaos Report 2021—Banks That Put Profit before the Planet [55] [56] [57]

Climate chaos is here and now. The World Meteorological Organization reports that disasters related to extreme weather have surged fivefold over the past fifty years. The main cause of the climate crisis is persistent use of dirty fossil fuels. Unfortunately, banks have a played a fundamental role in funding this dirty business, with fossil fuels receiving $5.9 trillion in subsidies every year. According to the International Energy Agency, net zero can only be achieved with no new oil or natural gas fields. Despite big banks being a part of the Net Zero Banking Alliance, subsidies and funding for fossil fuels continue on a scale of millions of dollars. According to a BBC article, banks, including HSBC, Barclays, and Deutsche Bank, are still backing new oil and gas despite being part of a green banking group.

An independent review by the Rainforest Action Network led to the release of the Banking on Climate Chaos Report 2021, which highlights that the world's sixty largest commercial and investment banks poured a total of $3.8 trillion into fossil fuels between 2016 and 2020. The top fossil-fuel-funding banks are JPMorgan Chase, Citigroup, Wells Fargo, Bank of America, the Royal Bank of Canada, and MUFG. Barclays is the worst offender in Europe, and the Bank of China is the worst in China. This encourages banks to reassess their financing policies for key fossil fuel sectors to impact the code red for business. Policies twenty-first-century leaders should adopt now with urgency include the following:

- Prohibit all financing for all fossil fuel expansion projects and for all companies expanding fossil fuel extraction and infrastructure along the whole value chain.
- Commit to measure, disclose, and set targets to zero out the absolute climate impact of overall financing activities on a +1.5°C-aligned scale by 2025.
- Commit to phase out all financing for fossil fuel extraction, combustion, and infrastructure, using an explicit time line.
- Fully respect all human rights, particularly the rights of Indigenous peoples, including their rights to their water and lands and the right to free, prior, and informed consent.

Fossil Fuel Companies Are Compromising the Climate Emergency / Code Red for Humanity

The major offenders exacerbating the climate emergency are fossil-fuel-exporting nations or ones reluctant to transition away from fossil fuels as the main source of energy and electricity. The public and private fossil fuel companies enabling the code red for humanity and business are Saudi Aramco, Exxon Mobil, BP, Shell, Chevron, Adani Coal, Petro China, Gazprom, and Sinopec, and the banks funding them include JPMorgan, Wells Fargo, Bank of America, Citigroup, Barclays, Standard Chartered, HSBC, Sumitomo Bank, and Deutsche Bank, as reported in the Banking on Climate Chaos Report 2021. The governance module focuses on the banking sector by reviewing the compromise made by banks that committed to stop funding fossil fuels at the December 2015 Paris Agreement talks but have since that been funding the code for humanity with complete and utter disregard.

The INSEAD case study of April 2021 focused on the forward thinking and twenty-first-century mindset of the leaders of the European Investment Bank, who stopped funding all fossil fuel projects in 2021. We must make a significant investment in the planet within the next decade if we wish to make the rapid transition that is required.

The three key case studies that cover the compromise of the banking sector (according to the Banking on Climate Chaos Report 2021) highlight the exemplary commitment of the EIB (European Investment Bank) to be the world's climate bank and the leadership of Ana Botin of Santander Bank.

4.9.2 The European Investment Bank INSEAD Case Study [58] [59]

Werner Hoyer, Leader of EIB, the European Climate Bank

The EU set a long-term vision based on the Intergovernmental Panel on Climate Change (IPCC) AR6 report to be on a decarbonization path where emissions will peak by 2020 at the latest and where 70 percent to 80 percent of energy requirements will be met using renewable sources by 2030, and thereafter quickly phasing out fossil-fuel-driven automobiles. To lead this green transition, the EIB as the EU's climate bank has set policies to mobilize funds in line with the EU Green New Deal.

This set of climate policies cements the bank's position as the leading multilateral provider of climate finance worldwide, and it happened under the guidance of the twenty-first-century science-led leadership of EIB president Dr. Werner Hoyer. Dr. Hoyer started his tenure at EIB in 2012, when he immediately started mobilizing funds (from 2013 to 2018, he invested more than twenty-eight billion euros in renewable energy projects and more than sixty-five billion euros in energy efficiency and distribution, powering forty-five million households). He also formulated the EIB's climate strategy in 2015. In November 2019, Dr. Hoyer announced an updated EIB climate strategy to advance the Green New Deal. He said in a statement, "It was decided to make a quantum leap in its ambitions. We will stop financing fossil fuels in 2021, and we will launch the most ambitious climate investment strategy of any public financial institution anywhere." Under the updated strategy, EIB will support one trillion euros worth of climate action and environmentally sustainable investment in the decade 2021 to 2030.

Dr. Hoyer has shown what it means to embed sustainability into a corporate strategy and corporate decision-making. He has also demonstrated to the world the following twenty-first-century science-led leadership traits:

- an appreciation for mobilizing science as part of the business strategy, by putting an end to EIB's financing of fossil fuels
- an urgent need to make an impact on the code red for business, by making a commitment to mobilize €1 trillion between 2021 and 2030 in support of climate action
- a triple-bottom-line mindset, prioritizing the planet over profit
- a willingness to embed the Twenty-First-Century Board Leadership Model in the business's board/C-suite/operations agenda—see EIB's transformation since 2012.

Ana Botin, Banking on Climate Stewardship—Santander Bank [60] [61]

Santander is among the world's ten most sustainable banks as ranked by the 2021 Dow Jones Sustainability Indices. Santander has featured in the DJSI for the past twenty-one years and is always ranked among the top performers. The bank also ranked fourth on the 2021 *Fortune*'s Change the World list. Being one of the largest banks in the world, it has a clear sustainability-led strategy to be loyal to its one hundred forty million customers. At COP26, twenty-first-century science-led leader Ana Botin made an appeal to all business organizations to prioritize the planet over profit considering the climate emergency and code red for humanity and business. In terms of the bank, some of the targets to transition to this new green economy are as follows:

- ✓ The bank has increased its commitments to renewable energy and climate-friendlier financing by mobilizing €120 billion for green finance by 2025 and €220 billion by 2030.
- ✓ Globally, 57 percent of the energy Santander uses for electricity generation is renewable, while Germany, Spain, Portugal, and the UK already using 100 percent green energy. And the bank pledges to use 100 percent renewable energy for electricity by 2025.
- ✓ Santander will stop financing energy clients who earn more than 10 percent of their revenues from coal by 2030.
- ✓ At COP26, Santander invested more than eleven billion US dollars in green energy projects and is also part of the Glasgow Financial Alliance for Net Zero, a group of four hundred fifty financial institutions committed to a net-zero goal, coal divestment, and climate reporting standards for the industry.

Ana Botin has shown what it means for the financial sector to be a first mover and make an impact on the climate emergency. She has demonstrated to the world the following twenty-first-century science-led leadership traits:

- • an appreciation for mobilizing science as part of the business strategy, by putting an end to the financing of coal power

- an urgent need to make an impact on the code red for business, by ensuring Santander uses 100 percent renewable energy for its electricity needs
- a triple-bottom-line mindset, prioritizing the planet over profit
- a willingness to embed the Twenty-First-Century Board Leadership Model in the business's board/C-suite/operations agenda, by investing more than eleven billion US dollars in green energy projects as of 2021.

4.9.3 Investment Required to Fight the Code Red for Humanity and Business

Is there a global shortage of funds to fight the code red for humanity and business? No. There is a lack of twenty-first-century leaders who are science led and have a sustainability mindset with an abundance of funds to channel given the required leadership.

On a planet where we spent US$5.9 trillion in 2021 to subsidize the cause of the climate emergency, that is, fossil fuels, and where we spend more than US$5.0 trillion per annum on defense, we find that the US$6.3 trillion to US$6.9 trillion per annum to accelerate the energy transition in order to save the planet from the code red for humanity and unprecedented economic devastation of between US$54 trillion and US$67 trillion when we cross the +1.5°C to +3.7°C line between 2025 and 2040 is less than 10 percent of estimated GDP.

Per the Organisation for Economic Co-operation and Development report of 2017:

- US$6.3 trillion will be invested each year until 2030 to meet the UN Sustainable Development Goals.
- US$6.9 trillion will be invested each year until 2030 to meet the terms of the December 2015 Paris Agreement.

With twenty-first-century leaders who understand the need to prioritize the planet over profit, we might see a combination of eliminating fossil fuel subsidies (with fossil fuels being the cause of the climate emergency) and reducing global defense expenditure, which could be the solution to fighting the code red for humanity and the business.

If we don't install twenty-first-century leaders who are ready to invest the required funds between 2022 and 2030 to secure the planet beyond 2030 for future generations, then we will simply be consenting to a continuance of the thirty-five years of climate action failure despite the fact that we now have the technologies required to drive the energy transition from fossil fuels to renewable energy.

Chapter 4—Twenty-First-Century Implications and Strategic Actions

Chapter 4 is the essence of *Twenty-First-Century Leadership to Fight the Code Red for Business* as in it we reviewed the Twenty-First-Century Board Leadership Model, which is the basis for addressing the code red for business. Understanding the facts, the reality, and the science behind the three emergencies and three disruptors is key.

1) Ceasing to use pollution-causing fossil fuels for energy, electricity, and transportation is no longer an option but is a first twenty-first-century imperative. Every nation, business, and organization should prioritize the move to sustainable renewable energy and transportation.

2) The sustainable energy transition must happen by 2025, or 2030 at the very latest, to prevent catastrophic global warming from devastating the planet.

3) Any leader of a nation, business, or organization who is science led will know why making commitments to end reliance on fossil fuels after 2030 is of no significance to undoing the climate emergency / code red for humanity.

4) Twenty-first-century science-led leaders have demonstrated that transitioning away from using fossil fuels and transitioning to 100 percent renewable energy is achievable, as shown by the leaders of Ørsted and South Australia.

5) The emergence of twenty-first-century science-led leaders at Alphabet/Google, Tesla, Unilever, Patagonia, EIB, Santander Bank, and Dilmah confirms that a new cadre of leaders is ascending.

6) Measuring carbon footprints of nations, businesses, and organizations needs to be done annually to establish the impact these entities are making on the planet and to take decisive steps at Scope 1, 2, and 3 to reduce that impact.

7) Climate emergency response strategies need to be developed and implemented to strategically reduce the carbon footprint and move to carbon-positivity with urgency.

8) Carbon offsetting is *not* a recommended strategy as it does not address the reality of reducing carbon emissions directly.

9) Global supply chains need to urgently reinvest and reinvent themselves to move from outsourcing and offshoring to near-sourcing and insourcing in order to reduce their carbon footprints. All twenty-first-century leaders will need to put the planet and people above profit.

10) The urgent need to replace all nineteenth-century science-denying nonleaders with science-led twenty-first-century leaders is an imperative, one that could determine the survival of humanity.

11) The economic emergency is mostly a result of ignoring the climate emergency and the health and social emergency. However, economic decisions that directly impact gross margins, investments, and working capital also contribute to the economic emergency.

12) Addressing the key challenges that cause the health and social emergency—deforestation, air pollution, water scarcity, plastic pollution, toxic industrial pollution, and food waste—should be the focus of every twenty-first-century leader. The responsibility of every organization is to identify both direct and indirect impacts its business has on each of these areas and to develop a strategy to eliminate or minimize those impacts.

13) The economic emergency requires us to review our business models and supply chains to address the possible impacts of the climate emergency and the health and social emergency.

14) The three disruptors need to be addressed from a strategic perspective, and a response needs to be initiated after analyzing every single possible disruptor—technology, geopolitics, and governance—to determine how it could make an impact on your business, either directly or indirectly.

15) The energy transition of the Southern Hemisphere nations should be funded by the Northern Hemisphere nations by doing the following:

   a) Rerouting existing subsidies for pollution-causing fossil fuels, with subsidies amounting to US$5.9 trillion in 2021. This is a no-brainer as investing any money in the very cause of the climate emergency—as is done by nineteenth-century science-denying leaders who persist with pollution-causing fossil fuels—is a ridiculous prospect.

   b) Reducing defense expenditure, which is another US$5 trillion to fight a world war that will never be fought, with the only beneficiary being the defense industry to fight the code red for humanity.

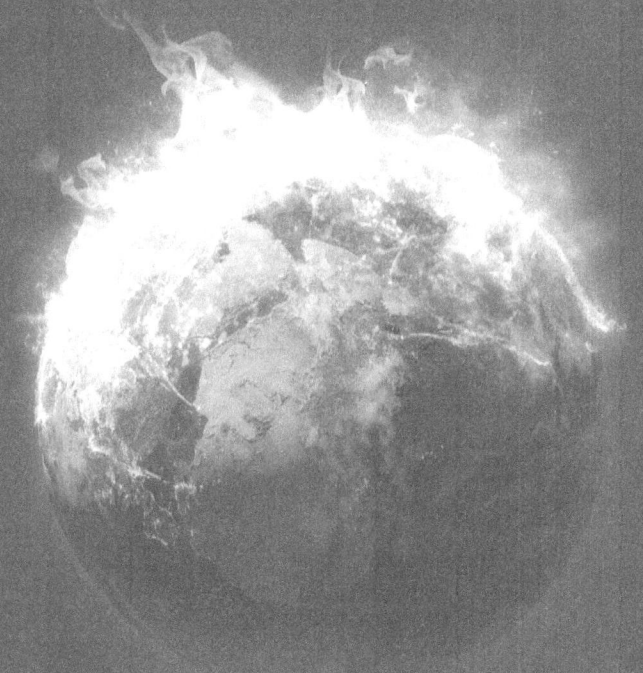

CHAPTER 5

BUILDING A TWENTY-FIRST-CENTURY STRATEGY TEMPLATE

CHAPTER 5

BUILDING A TWENTY-FIRST-CENTURY STRATEGY TEMPLATE

One of the unique benefits of attending a Twenty-First-Century Board Leadership Model MasterClass[1] is that one is able to see the step-by-step approach to developing the Twenty-First-Century Strategy template after each module. It also has the benefit of offering input from the peers and the faculty.

Every attendee of the MasterClass will be challenged, inspired, and guided to develop a company-specific twenty-first-century strategy with a comprehensive template much like the one shared in this chapter. Are you inspired to be a twenty-first-century leader? In this chapter, I will take you through the key elements of the strategy template and guide you on creating your own company-specific strategy.

A twenty-first-century leader is a person who demonstrates the following four unique strategic traits in his or her behavior:

1) an appreciation for mobilizing science as part of the business strategy
2) an urgent need to make an impact on the climate emergency / code red for business
3) a sustainability mindset, prioritizing the planet and people over profit.
4) a willingness to embed the Twenty-First-Century Board Leadership Model in the business's board/C-suite/operations agenda.

The twenty-first-century leader/strategist will be educated (with a functional area of expertise in business) and conversant in the required skills, experience, exposure and will have a global track record of having worked in at least two geographic regions. However, this alone is inadequate to perform in the twenty-first century. Such leaders need to supplement their knowledge with the Twenty-First-Century Board Leadership Model and be able to competently, professionally, analytically, and strategically develop a Twenty-First-Century Strategy Template.

The twenty-first-century strategist will need to first understand the potential impacts, both in the short term and in the long term, of each of the three emergencies and each of the three disruptions. Thereafter, the twenty-first-century leader will create a strategy that minimizes the impacts and takes strategic advantage of the trends and opportunities they present to the organization.

Having developed a Twenty-First-Century Strategy Template, twenty-first-century leaders will align it with the board, C-suite, and operations teams to ensure their buy-in and gain their input to implement it with urgency. Every board, C-suite, and operations team member needs to subscribe to the twenty-first-century science-led thinking and sustainability mindset.

The leaders will then evaluate and improve the template with input from a board strategy advisory committee made up of people who have expertise beyond the board. As mentioned earlier, no board has within itself the skills and expertise to address the six modules, and therefore the board needs to acknowledge that gap and supplement its deliberations with external input. In the case of the climate emergency, including technological and geopolitical disruptions, expert advice is necessary.

In chapters 1 to 4, the content highlighted the following:

➢ the reality facing humanity and businesses as the planet heats up
➢ the inaction of science-denying nineteenth-century leaders over the past thirty-five years (1987 to 2022) to transition away from pollution-causing fossil fuels
➢ the Twenty-First-Century Board Leadership Model and its six key modules that need urgent attention and response.

In chapter 5 we have the Twenty-First-Century Strategy Template that provides a step-by-step approach to developing a focused strategy for the business, which in turn can leverage all four traits of a twenty-first-century leader to move from knowledge to urgent action. Following is the Twenty-First-Century Board Leadership Model–based strategy template that is currently an essential part of the MasterClass that is being run in Luxembourg, Ireland, the United Kingdom, and Sri Lanka.

Before we go into creating the Twenty-First-Century Strategy Template, you need to assess the attitude and knowledge of your current leadership. The six-step approach is recommended for best results.

Module 1. Introduction and Review of Board Expertise

At this step, you need to challenge yourself to establish whether or not you have twenty-first-century leaders in place in your board, C-suite, and leadership team. You can establish this by reviewing the current incumbents of key positions and evaluate if they are twenty-first- or nineteenth-century leaders. Postevaluation, you could take one of the following actions:

1) Develop urgent supplementary programs in the event there are any gaps in knowledge, to ensure all board members and C-suite leaders are 100 percent up to speed.
2) Identify twenty-first-century leaders and induct them into board and board advisory positions to guide the organization.
3) In the future, ensure you identify and hire the right twenty-first-century talent for the organization.

It is also important to identify the top two emergencies and top two disruptors that can make a direct impact on the current business model. A strategic assessment must be done on the three emergencies (climate, social and health, and economic), and from there, the no. 1 emergency that needs an urgent response if the organization is to survive must be selected. Similarly, a strategic assessment must be carried out on the three disruptors (technology, geopolitics, and governance) to identify the no. 1 disruption that needs attention. The first step and the first module is to help focus the business/corporate response. The point must be made that while attention, resources, and focus are needed to respond to the no. 1 emergency and the no. 1 disruptor, one must not ignore the others, as they need to be addressed too.

Twenty-First-Century Board Leadership Model MasterClass

Twenty-First-Century Strategy Template

By completing a strategy template, the business opens an opportunity to create a company-specific twenty-first-century strategy for review by the board.

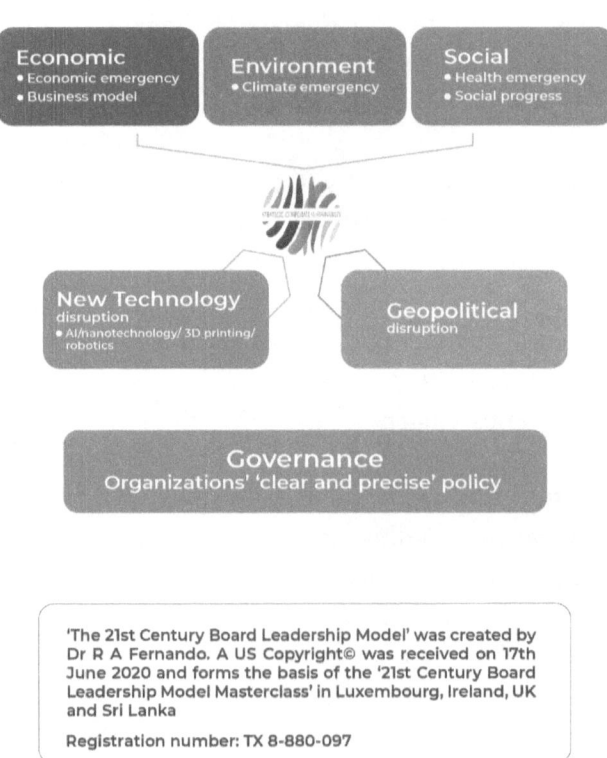

21st Century Board Leadership Model©

'The 21st Century Board Leadership Model' was created by Dr R A Fernando. A US Copyright© was received on 17th June 2020 and forms the basis of the '21st Century Board Leadership Model Masterclass' in Luxembourg, Ireland, UK and Sri Lanka

Registration number: TX 8-880-097

- From the three emergencies, you must identify which of them require urgent board and operations action to adapt. All business leaders must have an urgent need to solve the climate emergency by 2025.
- From the three disruptors, you must determine which require strategic advice and planned actions. Next, select the two that will make a direct impact on the business, then develop the corporate response strategy.

Step 1. Board Review

Module 1. Introduction and Review of Board Expertise

Board evaluation of the four imperatives needed to be a twenty-first-century leader

	Unaware	Early adopter	Focused	Expert	Role model
An appreciation for mobilizing science as part of the business strategy	☐	☐	☐	☐	☐
An urgent need to make an impact on the code red for business by 2025	☐	☐	☐	☐	☐
A triple-bottom-line mindset (prioritizing the planet and people over profit)	☐	☐	☐	☐	☐
A willingness to embed the Twenty-First-Century Board Leadership Model in the business's board/C-suite/operations agenda	☐	☐	☐	☐	☐

Do the board members have a twenty-first-century mindset? Identify the top two emergencies or areas where the company needs to supplement its internal strengths with external expertise. (Complete the following chart.)

Emergencies	Yes	No	Identify top two only
Climate (planet)	☐	☐	
Health/social (people)	☐	☐	
Economic (profit)	☐	☐	
Disruptors	**Yes**	**No**	
Technology	☐	☐	
Geopolitics	☐	☐	
Governance	☐	☐	

Step 2. Response to Emergencies

Complete Modules 2, 3, 4: Climate, Health and Social, and Economic Emergencies

This is the most rewarding and crucial step in your journey to be a twenty-first-century leader who both is science-led and has a sustainability mindset. In this step, we go into detail to identify the key challenges and the possible strategies that the business can launch to respond to all three emergencies in detail, giving priority to the two identified previously. We have identified the focus on the UN's Sustainable Development Goals should guide every strategy. This is a key exercise that will focus the organization's attention on what needs to be done as twenty-first-century leaders who need to ensure the business/organization is part of the solution to the twenty-first-century challenges instead of being a part of the problem.

In the case of both emergencies and disruptors, it is key that you identify the likely impacts in both the short term and the long term and thereafter make an informed decision as to which ones are most likely to make a direct impact on your business.

As you invest time in building a twenty-first-century strategy for your organization using the template, considering some key insights and advice as you complete modules 2, 3, and 4 and respond to the three key emergencies.

The need to respond with urgency to the three emergencies after considering the scientific advice is not an option. The word *emergency* has unfortunately been ignored by nineteenth-century science-denying leaders. You no longer belong to that group, who have let the planet deteriorate to the point that we are now facing a code red for business.

As you respond to each emergency, consider the organization you lead and ask the question, In terms of pollution, is my business part of the problem or part of the enlightened twenty-first-century science-led solution?

- **Climate emergency**
  The cause of the climate emergency has been plain for the past thirty-five years. It's the burning of fossil fuels—coal, petroleum, diesel, and natural gas. The transition from pollution-causing fossil fuels to renewable energy sources must be achieved by 2025 ideally or by 2030 at the latest. Building sustainable supply chains is the second key focus. Compromising either of these efforts is a recipe for disaster.

- **Health and social emergency**
  The need to identify the key areas of compromise that are contributing to the global health and social emergency, either directly or indirectly, is the first step. The key issues highlighted are by no means a comprehensive list. But it's a start for a business that wishes to be part of the solution. Pandemics, air pollution, deforestation, water scarcity, plastic pollution, and industrial pollution are among the key topics addressed. The strategy developed must directly affect every single are the business affects either directly or indirectly.

- **Economic emergency**
  Every economic emergency is as a result of ignoring and not acting on solving the climate emergency and the health and social emergency. The major focus in this module is to ensure your business has a sustainable twenty-first-century business model and that it drives sustainable innovation to significantly reduce or entirely eliminate the carbon footprint of the organization in terms of Scope 1, 2, and 3 emissions.

Modules 2, 3, and 4—Climate, Health and Social, and Economic Emergencies

Step 3. Carbon Footprint Exercise

Twenty-First-Century Model Risk Assessment

Summarize one risk, one opportunity, or one action for all three emergencies (modules 1, 2, 3, and 4) and all three disruptors (modules 5, 6, and 7).

Emergencies			
	Briefly summarize two risks	Briefly summarize one opportunity	Actions to take
Climate emergency	Risk 1:	Opportunity 1:	Action 1:
Health emergency			
Social emergency			
Economic emergency			
Disruptors			
New technology disruption	Risk 1:	Opportunity 1:	Action 1:
Geopolitical disruption			
Governance impact (placing profit before the planet)			

Twenty-First-Century Model–Based Corporate Strategy

In each case, the strategy will be led by a select priority UNSDG. I have identified five key UNSDGs that should be part of every strategy and are nonnegotiable. These are UNSDGs 7, 9, 12, 6, and 15.

Step 4. Focused UNSDG Strategy

NB: Companies that offer financial services should invest in the following:

What clean-energy-related initiatives/goals can the organization commit to with urgency?

Twenty-First-Century Business Model Impacting UNSDG 7
(Complete the following chart.)

End reliance on pollution-causing fossil fuels with urgency (UNSDG 7)

Vision 2025 New energy mix	
New company fleet mix and logistics (fossil fuel vehicles, electric vehicles, hybrid vehicles)	
Target milestones for UNSDG 7	
Action plan	
Dates	
Who is responsible	

Twenty-First-Century Model Supply Chain Strategy (UNSDG 12)

In-house supply chain (Scope 1 and Scope 2)	
Briefly describe your organization's twenty-first-century supply chain strategy (e.g., sustainable sourcing, manufacturing, warehousing and distribution, end consumer consumption, life cycle analysis).	
Where are you focused for 80 percent of your business (e.g., product and/or service)?	

What changes are you making toward sustainable production and consumption?	
Outsource supply chain (Scope 3)	
What are the key supply chains or supply chain investments that could be disrupted by the climate emergency?	
Important considerations	
Near-shoring?	
Inshoring?	
Vision 2025	
Milestones	
Action plan and dates	
Who is responsible	

Managing Future Pandemics (UNSDG 3)

Code Red Strategy

(Complete the following chart.)

Here we are speaking of a strategy (to be implemented within forty-eight hours) to minimize transmission of illness during any future pandemics at both company level and supply chain level.

In-company	
Key milestones (in-company)	
Action plan and dates	
Who is responsible	
KPIs	
External suppliers	
Key milestones (external suppliers)	
Action plan and dates	
Who is responsible	
KPIs	

Identify one key new policy/strategy to create a sustainability mindset among employees, the community, and society and to enable sustainable lifestyles and reduce inequality. (Complete the following chart.)

Sustainability ≠ CSR

Employees	
Community	
Society	

Twenty-First-Century Business Model for Sustainable Innovation and Infrastructure, Impacting UNSDG 9

Identify a priority area of focus for twenty-first-century sustainable innovation strategy (refer to Module 5—Technology Disruption)	
Milestones, action plan, and who is responsible	

Twenty-First-Century Business Model Impacting Environment/Planetary Boundaries—UNSDGs 6, 14, and 15

Identify a priority area of focus for the twenty-first-century sustainable innovation strategy	
Milestones and action plan	

Step 5. Disruptor Response Strategy

Complete modules 5, 6, and 7: the disruptors—technology, geopolitical, and governance.

Having addressed the emergencies and created a twenty-first-century business model to ensure the organization focuses on being part of the solution to the code red for business, we now move to the second half of the model, where we focus on the three disruptors. It must be noted that none of the six modules covered in the Twenty-First-Century Leadership Model are covered in traditional MBA programs or board governance programs.

A twenty-first-century leader initially must be conversant in and up to speed with the three disruptors. This will require investing time and resources so that the leader may be updated on how the possible disruption caused by any one of the three disruptors could affect business as usual.

It is key that a twenty-first-century leader take a strategic approach to addressing the three disruptors:

- **Technology disruption**
  The world of science-led technology advancement is moving at a rapid clip. Just keeping pace with the new technology, which can have a direct or indirect impact on the current business, is in itself a challenge and requires a commitment to having the talent and strategic advisory expertise necessary to update the CEO, board, and C-suite on the possible disruptions in the medium to long term. Among the game-changing technology developments are smart grids, advances in solar panels (perovskite) plus battery storage, artificial intelligence, precision agriculture, nanotechnology, robotics, 3D printing, electric rail, and sustainable transportation.

- **Geopolitical disruption**
  Every business has a nation of origin, and most nations have geopolitical issues and trade barriers that directly affect their ability to operate freely in most areas. The twenty-first-century leader is fully apprised, understands the global geopolitical trends that could disrupt the business, and will not be caught unawares of what is happening in this space—once again, a subject *not* discussed in most MBA programs. Today, there are three key geopolitical actors that influence most global

geopolitics: China with its One Belt One Road strategy and its focus on unifying greater China; the USA, which is increasingly inward-looking and divided; and Russia, whose focus is moving back to the days of the Soviet Union and influencing the European Union. An understanding of global debt and water scarcity will influence geopolitical decisions also needs to be taken into consideration.

- **Governance disruption**
  The twenty-first-century leader and strategist is someone who is led by science and, as a result, will not compromise the planet. The new paradigm for twenty-first-century governance is to be a leader who prioritizes the planet and people over profit. Nineteenth-century science-denying leaders have compromised the planet for the past thirty-five years to such an extent that it has given rise to the code red for business and humanity. This is the reality.

The twenty-first-century leader with a commitment to being led by science understands that he or she cannot eliminate risk, but must at all times evaluate all the options available to minimize the risk to and impact on both the planet and people. In other words, such a leader must focus on environmental sustainability and social sustainability. Today, the only determining factor is which option will generate the greatest profit. This needs to change. The global compromise to subsidize fossil fuels, the cause of the climate emergency / code red for humanity, to the tune of US$5.9 trillion per annum is unacceptable and does not align with the science. The investment by nineteenth-century science-denying leaders in global banks and financial institutions needs to be halted, with this money being channeled to accelerate the transition to renewable energy.

The twenty-first-century leader/strategist understands the need for proactive action to counter the three disruptors.

Module 5—Technology Disruption

Which new technologies could disrupt your business, and which ones should you pursue?

Technology	
Vision 2025	
Milestones	
Action plan and Dates	
Who is responsible	

Module 6—Geopolitical Disruption

Which one key geopolitical challenge or opportunity could disrupt your current business model?

Risk What are the potential geopolitical risks that can disrupt the business model (or funding), supply chain, and strategy? Examples include the US-China trade war, Huawei, Brexit, and the water wars. **Opportunity** Which opportunities should you leverage? Examples include Brexit and China's One Belt One Road initiative.

Risk/opportunity management strategy and milestones	
Action plan and dates	
Who is responsible	
KPIs	

Module 7—Governance: Planet before Profit

This module deals with business models, policies, and strategies that compromise the governance principle and how they should be changed. (Complete the following table.)

Identify one current area of compromise where the company prioritizes profit over the planet and people, and develop new policies to ensure zero compromise both from the company and in the supply chain (i.e., prioritizing shareholder over stakeholder value).	Action plan to change those identified. (Please include proposed dates and who is responsible.)
Current area 1:	Action plan to change:

Step 6. Formulate a Twenty-First-Century Future Board Strategy Advisory Committee

No board, C-suite, or leadership team on the planet can have *all* the relevant expertise concentrated in the organization. Even if one did have all the technical expertise, it is unlikely that it would also have the knowledge, track record, and knowledge of current market trends, or an awareness of likely future breakthroughs.

To be a twenty-first-century-ready business that is able to respond with clarity, focus, and expertise to the six modules of the Twenty-First-Century Board Leadership Model, any business must have on its Twenty-First-Century Future Board Strategy Advisory Committee a number of global experts on the key subject areas. As to the priority areas where a major knowledge gap exits, it is clear that a majority of organizations do not have the knowledge and expertise to do the following:

1) Respond to the climate emergency, as it's an evolving, dynamic, and fluid situation where new knowledge and the latest technological advances must be adopted.
2) Respond to the health and social emergency.
3) Respond to a technology disruption, for which strategic futurist input is crucial.
4) Respond to a geopolitical disruption with a grasp of the realities that are unfolding in this complex space.

Chapter 5. Twenty-First-Century Strategy Template—Implications and Actions

1) Creating a Twenty-First-Century Strategy Template is the key step one must take to put into practice all the key learning and, most importantly, operationalize the six key modules that cover the three emergencies and the three disruptors for the organization.

2) Step 1 and step 2 are about self-evaluation and analysis of one's twenty-first-century leadership preparedness and what steps need to be taken to move forward.

3) Step 3 is about making a high-integrity realistic evaluation of the competence of the board and C-suite to address the twenty-first-century challenges. The objective is to either develop the current leaders to be twenty-first-century-ready or replace them with people who are. One could also have a hybrid solution where one develops the current leaders and injects expertise in the form of a board advisory committee to fill in the gaps.

4) Step 4 is about addressing the three modules covering the emergencies and to do so with urgency and commitment, mobilizing all available resources to ensure the business is part of the solution and not a part of the problem.

5) Step 5 is about addressing the three modules covering the disruptions and to do so with strategic acumen so as to ensure the business is *not* caught unawares or unprepared for any of these by having a documented strategy.

6) Step 6 is about formulating a board advisory strategy based on the gap analysis to augment the expertise and knowledge required to make twenty-first-century strategy decisions.

7) Step 7 is once again about self-evaluation of the personal journey and progress one has made to move from being a nineteenth-century science-denying leader to being a twenty-first-century science-led leader. Consider investing in the Twenty-First-Century Board Leadership Model MasterClass to become a more relevant twenty-first-century leader.

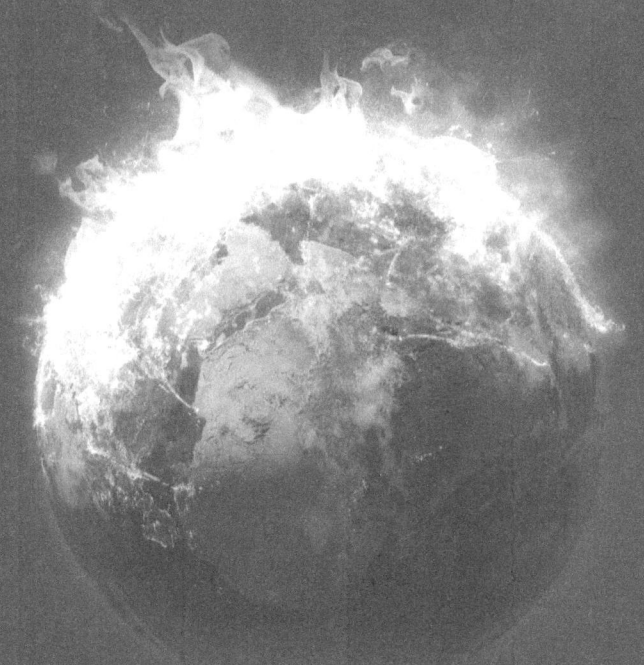

CHAPTER 6

ARE YOU READY TO FIGHT THE CODE RED FOR BUSINESS?

CHAPTER 6

ARE YOU READY TO FIGHT THE CODE RED FOR BUSINESS?

Today, as you come to the end of *Twenty-First-Century Leadership to Fight the Code Red for Business*, you can decide on one of two ways to proceed:

1) Ignore the code red for humanity and business, as a majority of science-denying nonleaders of nations have done for the past thirty-five years, and continue with business as usual with pollution-causing fossil fuels.
2) Commit to enlisting as a twenty-first-century science-led leader to fight the code red for humanity and business with urgency.

In each of the chapters, we explored a key twenty-first-century challenge and the crisis that will result if we don't act with urgency to fight the code red for humanity and business:

- Chapter 1. Is There a Code Red for Humanity and Business?
- Chapter 2. Thirty-Five Years of Science Denial—a Failure of Climate Action from 1987 to 2022

We explored the reality that has led to the climate emergency and to the code red for humanity and business. It is clear that for the past thirty-five years since the Bruntland report was presented, world leaders, business leaders, leaders of the UN, and leaders of all organizations with the influence to act on the available science have denied the science and failed to act with focus and urgency to end our reliance on pollution-causing fossil fuels. This has compromised the planet and all future generations. The inaction has led to climate action failure, with the climate emergency now (in 2022) ranked as the no. 1 global risk by the World Economic Forum.

Implication

If we continue this global trend, which hit a crescendo of inaction and denial at COP26 in Glasgow, we will be in trouble. Despite the World Meteorological Organization's 2021 report and the Intergovernmental Panel on Climate Change's 2021 report, which both discuss the likelihood that the planet will hit the tipping point of +1.5°C by 2025[1] or 2030 and even surpass it, with global temperatures rising by 2.2°C–2.5°C, world nonleaders decided not to take decisive climate action to transition away from using pollution- and climate-emergency-causing fossil fuels for the twenty-sixth time.

Action Required

It is becoming almost inevitable that we will not be able to avoid this increase in global temperatures unless we cease using fossil fuels by 2025 or, worst case, by 2030. The projected devastation to the global economy, to the tune of US$54 trillion to US$67 trillion,[2] seems to be the new reality we are facing in the absence of twenty-first-century leadership.

We need twenty-first-century leaders to step up to the plate and drive four urgent actions:

1) Cease using pollution-causing fossil fuels by 2025–2030.
2) Halt all subsidies for fossil fuels by 2022–2025.

3) Halt deforestation by 2022–2025.
4) Transition to 100 percent sustainable energy and transportation by 2030.

Do we have twenty-first-century leaders ready to do these things?

- Chapter 3. Twenty-First-Century Leadership?

In chapter 3, we discussed the characteristics of a twenty-first-century leader:

1. an appreciation for mobilizing science as part of the business strategy
2. an urgent need to make an impact on the code red for humanity and business by 2025
3. a triple-bottom-line mindset, prioritizing the planet and people over profit
4. a willingness to embed the Twenty-First-Century Board Leadership Model in the business's board/C-suite/operations agenda.

It is key that each leader in a position of influence be a steward of the planet and get up to speed to move from being a nineteenth-century, science-denying, climate-action-failure-enabling nonleader to being a twenty-first-century science-led leader who champions the Twenty-First-Century Board Leadership Model and drives the twenty-first-century strategy with urgency.

- Chapter 4. The Twenty-First-Century Board Leadership Model[3]

In chapter 4, we presented the Twenty-First-Century Board Leadership Model. The challenge to all those in any positions of leadership is to understand the reality that is unfolding and how it will directly and indirectly make an impact on the business or the nation.

21st Century Board Leadership Model$^©$

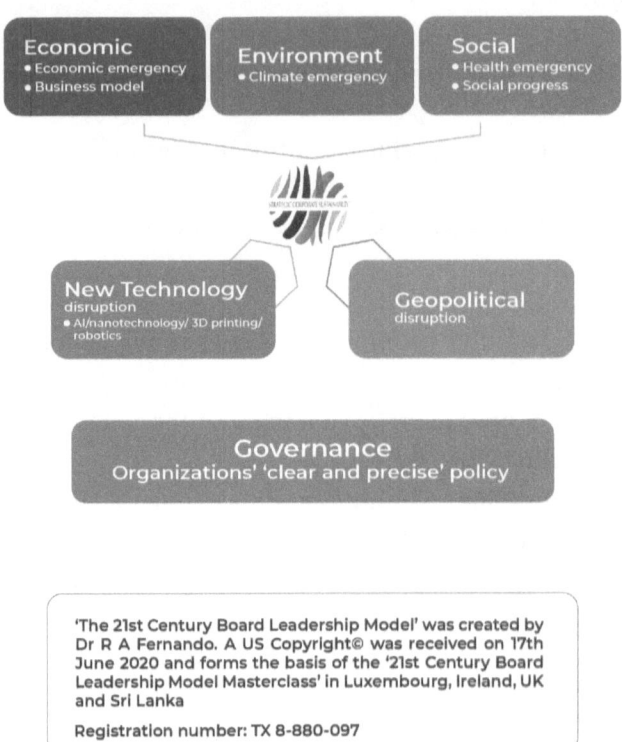

'The 21st Century Board Leadership Model' was created by Dr R A Fernando. A US Copyright© was received on 17th June 2020 and forms the basis of the '21st Century Board Leadership Model Masterclass' in Luxembourg, Ireland, UK and Sri Lanka

Registration number: TX 8-880-097

Urgent and decisive climate action and leadership is needed to combat the three emergencies, and strategic acumen is required to deal with the three disruptors.

Twenty-first-century leadership involves having a comprehensive understanding of each of the six modules and how each of them will affect the current business model and feeling an urgent need to respond to each of them.

○   4.0. The Emergencies

Appreciate the need for urgency to make an impact on the code red for humanity / climate emergency as a twenty-first-century leader, versus having the nineteenth-century, science-denying, profit-at-any-cost-to-the-planet attitude that is prevalent among most people in positions of influence.

- 4.1. The Climate Emergency and Extreme Weather

Ending our reliance on pollution-causing fossil fuels and enabling the urgent energy transition is the focus of a twenty-first-century science-led leader. The focus of the chapter is to do this by 2025 ideally, or 2030 at the latest, and to reset all supply chains to be sustainable. The case studies of Ørsted (Denmark) and South Australia confirm that this can be done.

- 4.2. The Health and Social Emergency

The health and social emergency requires a deep understanding of the current business model and how every one of the key health challenges identified is currently affected by it. Once that analysis is done, the focus of the chapter is to create a response that ensures the company is part of the solution and does not continue to be part of the problem. Key issues such as global debt, water scarcity, and pandemics, and focused issues such as waste, pollution, deforestation, and single-use plastics, need to be addressed. Being an organization that enables and encourages all its employees and families to move toward living sustainable lifestyles is a key focus. The INSEAD case studies of Singapore, Dilmah, and CDB Advance confirm that this can be achieved.

- 4.3. The Economic Emergency

Twenty-first-century leaders understand that ignoring the climate emergency and the health and social emergency is what causes nations and businesses to have an economic emergency. The need to have a sustainable business model and a focus on driving sustainable innovation is the major focus of the chapter, with case studies of those who have done so successfully. The case studies on Patagonia, Unilever, and Tesla confirm that with twenty-first-century leaders such as Yvon Chouinard, Paul Polman, and Elon Musk, nothing is impossible.

○ 4.4. The Disruptors

The twenty-first-century leader is someone who is strategic in terms of his or her actions and who anticipates all possible disruptors that can make an impact on the business in both the medium term and the long term, with advice from experts on each of the modules.

▪ 4.4.1. The Technology Disruption

The twenty-first-century leader has an up-to-date grasp of the global technology landscape and of major developments and advancements. An analysis of which of these technologies has the ability to disrupt the business needs to be done, and a strategy needs to be developed with expert advice on which technologies the company needs to pursue in order to be relevant in the twenty-first century. A board awareness of the new high-technology-driven power grids, perovskite solar panels plus battery storage, AI, precision agriculture, nanotechnology, graphene water desalination, robotics, and nano–bio is crucial as these are among the game-changing technologies that are now part of the global landscape.

▪ 4.4.2. The Geopolitical Disruption

Twenty-first-century leaders understand that in a complex and connected world, the best-laid corporate strategies can be disrupted by geopolitical alignments and tensions. It is crucial to have an up-to-date understanding of the geopolitical landscape in nations and markets the business depends on. It is also important to have in place strategic initiatives and fallback strategies that anticipate possible disruptions and strategic opportunities. A new area of expertise is required if a business wishes to operate in the twenty-first century. The company needs to acquire or seek the advisory services of those who are well-informed of the impending political maneuvers that could impact the business or present an opportunity to grow. This will enable the business to preemptively manage the geopolitical disruption and ensure that company performance is not devastated by at least minimizing the risk—or being a first mover in the case of an opportunity. Most business schools do not offer courses in this subject as part of the curriculum. The US-China trade relations, the China One Belt One Road strategy, the EU Green New Deal, and Russian influence over the European Union are areas to be sensitive to.

- 4.4.3. The Governance Disruption

Twenty-first-century leaders know the base paradigm for twenty-first-century governance where every decision takes into account the likely scenarios that will impact the investment decision with the least amount of disruption. Ensuring every decision made prioritizes the planet and people over profit is the key concern of a twenty-first-century leader focused on a triple bottom line. The decisions being made will always be a trade-off between the strategy with the most detrimental impact and the one with the most minimal impact. The key is that the leader has done the due diligence to minimize the impact and to take strategic advantage of opportunities.

The Banking on Climate Chaos Report 2021 takes a close look at the global banking sector and confirms that a majority of the boards, CEOs, and C-suite leaders in this sector compromise the planet in exchange for profit. The irresponsible decision to invest in fossil fuels since the 2015 December Paris Agreement, with more than US$2 trillion to US$3 trillion invested in fossil fuel projects by banks such as JPMorgan Chase, Wells Fargo, Citigroup, HSBC, Barclays, Standard Chartered, Deutsch Bank, Sumitomo Bank, Mitsubishi Bank, and the Royal Bank of Canada, has shown that these banks lack governance in their decision-making.

The fact that these banks have invested in the fossil fuel projects of Saudi Aramco, Exxon Mobil, Chevron, Adani Coal, Shell, BP, et al., is evidence that nineteenth-century leaders in board positions are influencing these organizations. The INSEAD case on the European Investment Bank in Luxembourg is the benchmark for governance in the financial sector.

- Chapter 5—Building a Twenty-First-Century Strategy Template

Twenty-first-century leaders know that they need to move beyond the reality and develop a twenty-first-century strategy based on the provided template (which can be adapted to meet the organization's needs). The final piece of evidence providing that we have created a twenty-first-century leadership cadre will be when many organizations have revamped all their corporate strategies by incorporating the Twenty-First-Century Board Leadership Model.

- Chapter 6. Twenty-First-Century Leadership to Fight the Code Red for Business

The final chapter, which is a call to action, expresses the commitment to inspire a new cadre of twenty-first-century leaders to drive the new paradigm.

What is your decision? If it is to be a twenty-first-century science-led leader, then sign below to enlist and fight the code red for business today.

I, _____ [your name], will be a twenty-first-century science-led leader to fight the code red for business from _____ [date] onward with urgency in order to make an impact on the climate emergency.

_____

Signature

A number of exemplary twenty-first-century leaders are fighting the code red for business today. If you signed on the line, then you are in good company with a host of global leaders. Let's look at a few of them and see what their twenty-first-century leadership achievements are.

Twenty-First-Century Business Leaders

1) Elon Musk of SpaceX, Tesla, and Solar City is clearly the visionary leader of the century.

2) Yvon Chouinard, founder of Patagonia, set the global benchmark for twenty-first-century leadership and created the business model for Patagonia from its inception.

3) Henrik Poulsen of Ørsted (Denmark) proved to the world that fossil fuels have no place in the twenty-first century. He created Ørsted from Danish Oil and Gas—a lesson for all nineteenth-century-led fossil fuel companies.

4) Paul Polman set a new paradigm for global FMCG companies by resetting Unilever with its Unilever Sustainable Living Plan.

5) Ana Botin of Santander Bank set the pace for the banking sector.

6) Werner Hoyer of the European Investment Bank leads the world's largest multilateral financial institution. By ending funding for all fossil fuel projects, he has set the global standard for how a responsible financial institution should be led.

7) Sundar Pichai of Alphabet/Google fame has been setting the benchmark for all global Internet giants to follow with his 24/7 carbon-positive strategy, having already moved his company to 100 percent renewable energy.

Twenty-First-Century Global Influencers

1) Sir David Attenborough and Johan Rockström has been consistently waking the world up to the reality of the climate emergency.

2) Greta Thunberg has been the most vibrant and passionate advocate for making a positive impact on the climate emergency, consistently highlighting the need to heed the science and act with urgency as we exhaust the global carbon budget. A Nobel Prize is long overdue for this twenty-first-century leader who stands in proxy for all future generations.

3) John Elkington created the triple-bottom-line concept in 1997 to help all leaders have a strategic framework to create a sustainability strategy.

Twenty-First-Century Leadership Dilemma

Twentieth-century board leadership	Twenty-first-century board leadership
➢ Flawed business model ➢ Prioritizing profit over the planet and people ➢ Being part of the problem ☐ Shareholder value creation ☐ Financial/annual reports ☐ Carbon offsetting (a license to pollute?) ☐ Seventeen UNSDGs—marginal impacts ☐ A business model driven by fossil fuels	☐ Sustainable business model ➢ Prioritizing the planet and people over profit ☐ Being part of the solution ☐ Stakeholder value creation ☐ Integrated report ☐ Carbon-negative ☐ Focus on priority UNSDGs 7, 9, 12, 6, and 15 ☐ A business model driven by renewable energy

As we come to the end of *Twenty-First-Century Leadership to Fight the Code Red for Business*, we know what twenty-first-century leaders should focus on doing, which is the opposite of what nineteenth-/twentieth-century, business-as-usual, science-denying leaders have been doing. Today, you can be a part of the solution to the climate emergency and the code red for business by making a commitment to act with urgency within your sphere of influence to ensure all future generations have a planet to live on.

Begin the journey to becoming a twenty-first-century leader!

ENDNOTES

Chapter 1. Is There a Code Red for Humanity and Business?

[1] David Shukman, "2021 Climate: World at Risk of Hitting Temperature Limit Soon," BBC, May 27, 2021, https://www.bbc.com/news/science-environment-57261670.

[2] *Breaking Boundaries: The Science of Our Planet*, directed by Jon Clay (London: Silverback Films, 2021).

[3] *Don't Look Up*, directed by Adam McKay (New York: Netflix / Bluegrass Films, 2021).

[4] Greta Thunberg, Svante Thunberg, Malena Ernman, and Beata Ernman, *Our House Is on Fire: Scenes of a Family and a Planet in Crisis* (New York: Penguin, 2021).

[5] O. Hoegh-Guldberg, D. Jacob, M. Taylor, M. Bindi, S. Brown, I. Camilloni, A. Diedhiou, R. Djalante, K. L. Ebi, F. Engelbrecht, J. Guiot, Y. Hijioka, S. Mehrotra, A. Payne, S. I. Seneviratne, A. Thomas, R. Warren, and G. Zhou, "Chapter 3: Impacts of 1.5°C Global Warming on Natural and Human Systems" (2018), in "Global Warming of 1.5°C: An IPCC Special Report on the Impacts of Global Warming of 1.5°C above Pre-Industrial Levels and Related Global Greenhouse Gas Emission Pathways, in the Context of Strengthening the Global Response to the Threat of Climate Change, Sustainable Development, and Efforts to Eradicate Poverty," V. Masson-Delmotte, P. Zhai, H.-O. Pörtner, D. Roberts, J. Skea, P. R. Shukla, A. Pirani, W. Moufouma-Okia, C. Péan, R. Pidcock, S. Connors, J. B. R. Matthews, Y. Chen, X. Zhou, M. I. Gomis, E. Lonnoy, T. Maycock, M. Tignor, and T. Waterfield, eds., in press.

6   BBC News, "Life at 50°C: How to Make a Film in the Hottest Places on Earth," November 7, 2021, https://www.bbc.com/news/av/world-59186087.

7   *WMO Atlas of Mortality and Economic Losses from Weather, Climate, and Water Extremes (1970–2019)*, World Meteorological Organization (WMO), https://library.wmo.int/index.php?lvl=notice_display&id=21930#.YlMgRshBzIU.

8   M. Burke, S. Hsiang, and E. Miguel, "Global Non-Linear Effect of Temperature on Economic Production," *Nature* 527 (2015): 235–39, https://doi.org/10.1038/nature15725.

9   "Global Risks Report 2022," World Economic Forum, https://www.weforum.org/reports/global-risks-report-2022.

10   R. Lindsey, "Climate Change: Atmospheric Carbon Dioxide," National Oceanic and Atmospheric Administration, https://www.climate.gov/news-features/understanding-climate/climate-change-atmospheric-carbon-dioxide.

11   *WMO Atlas of Mortality.*

Chapter 2. Thirty-Five Years of Science Denial—a Failure of Climate Action from 1987 to 2022

1   "Net Zero by 2050—Analysis," IEA, https://www.iea.org/reports/net-zero-by-2050.

2   "Composite Graph of Atmospheric CO_2 at Mauna Loa Observatory, December 2021—Scripps Institution of Oceanography and NOAA Global Monitoring Laboratory," graphics and lead scientist, Ed Hawkins, National Center for Atmospheric Science, University of Reading.

3   World Commission on Environment and Development, *Our Common Future* (Oxford: Oxford University Press, 1987).

4   M. Fischetti, "We Are Living in a Climate Emergency, and We're Going to Say So," *Scientific American*, April 12, 2021, https://www.scientificamerican.com/article/we-are-living-in-a-climate-emergency-and-were-going-to-say-so/.

5   *Encyclopædia Britannica*, s.v. "Timeline of climate change," https://www.britannica.com/story/timeline-of-climate-change, accessed October 30, 2022.

6   R. Pardikar, "Global North Is Responsible for 92% of Excess Emissions," *Eos*, October 28, 2020, p. 101, https://doi.org/10.1029/2020EO150969.

7   "Kofi Annan on the UN Millennium Goals," Kofi Annan Foundation, https://www.kofiannanfoundation.org/in-the-news/kofi-annan-on-the-un-millennium-goals/#:~:text=In%202000%2C%20the%20international%20community,Annan%20provides%20a%20progress%20report.

8   "The Paris Agreement," UNFCC, https://unfccc.int/process-and-meetings/the-paris-agreement/the-paris-agreement.

9   "Emissions Gap Report 2021: The Heat Is On—a World of Climate Promises Not Yet Delivered, Executive Summary, Nairobi," United Nations Environment Programme, 2021.

10   "Climate Change: Vital Signs of the Planet," NASA, https://climate.nasa.gov/.

11   Florian Zandt, "Infographic: Humanity's Uneven CO_2 Footprint," Statista Infographics, December 16, 2021, https://www.statista.com/chart/26416/average-co%25E2%2582%2582-emissions-per-capita-in-selected-regions/.

12   IEA, "Net Zero by 2050—Analysis," https://www.iea.org/reports/net-zero-by-2050.

13   Kate Larsen, Hannah Pitt, Mikhail Grant, and Trevor Houser, "China's Greenhouse Gas Emissions Exceeded the Developed World for the First Time in 2019," Rhodium Group, May 6, 2021, https://rhg.com/research/chinas-emissions-surpass-developed-countries/.

14   D. Carrington, "Fossil Fuel Industry Gets Subsidies of $11 Million a Minute, IMF Finds," the *Guardian*, October 6, 2021, https://www.theguardian.com/environment/2021/oct/06/fossil-fuel-industry-subsidies-of-11m-dollars-a-minute-imf-finds.

15   "Fossil Fuel Subsidies," IMF, https://www.imf.org/en/Topics/climate-change/energy-subsidies.

16   F. Harvey, "COP26: World on Track for Disastrous Heating of More Than 2.4°C, Says Key Report," the *Guardian*, November 9, 2021, https://www.theguardian.com/environment/2021/nov/09/cop26-sets-course-for-disastrous-heating-of-more-than-24c-says-key-report.

17   "CAT Emissions Gap, 2021," Climate Action Tracker, https://climateactiontracker.org/global/cat-emissions-gaps/.

Chapter 3. Twenty-First-Century Leadership

[1]   R. A. Fernando, Twenty-First-Century Board Leadership Model, US copyright TX 8-880-097, 2020.

[2]   John Elkington, "25 Years Ago I Coined the Phrase 'Triple Bottom Line.' Here's Why It's Time to Rethink It," *Harvard Business Review*, June 25, 2018, https://hbr.org/2018/06/25-years-ago-i-coined-the-phrase-triple-bottom-line-heres-why-im-giving-up-on-it.

Chapter 4. Twenty-First-Century Board Leadership Model

[1]   *Global Risks Report 2022*, World Economic Forum, https://www.weforum.org/reports/global-risks-report-2022.

[2]   "Sustainability—Enabling Sustainable Growth," Ørsted, https://orsted.com/sustainability2020.

[3]   "Ørsted," in "Top 100 Companies 2022," *Sustainability*, https://sustainabilitymag.com/magazine/top-100-companies-january-2022.

[4]   "Case Study—Ørsted," Science Based Targets, https://sciencebasedtargets.org/companies-taking-action/case-studies/orsted.

[5]   "EIB Backs Ørsted with €500 Million Loan Agreement," *CSRWire*, November 5, 2021, https://www.csrwire.com/press_releases/730941-eib-backs-orsted-eu500-million-loan-agreement-boosts-green-energy.

[6]   Christer Tryggestad, "Ørsted's Renewable-Energy Transformation," an interview with Martin Neubert, McKinsey Sustainability, July 10, 2020, https://www.mckinsey.com/business-functions/sustainability/our-insights/orsteds-renewable-energy-transformation#.

[7]   T. Barsoe and S. Jacobsen, "Ørsted Plans $57 Billion Drive to Be No. 1 in Green Energy," Reuters, June 2, 2021, https://www.reuters.com/article/us-orsted-outlook-idCAKCN2DE0X6.

[8]   "South Australian Government Climate Change Action Plan 2021–2025," Government of South Australia, 2020, https://cdn.environment.sa.gov.au/environment/docs/climate-change-action-plan-2021-2025.pdf.

[9]   E. Musci, "Leading the Green Economy. Department for Energy and Mining," Government of South Australia, 2022, https://www.energymining.sa.gov.au/growth_and_low_carbon/leading_the_green_economy.

10 Sean Fleming, "7 Renewable Energy Lessons from South Australia," World Economic Forum, June 2021, https://www.weforum.org/agenda/2021/06/renewable-energy-south-australia-climate-change/.

11 "All In on Renewable Energy: Jay Weatherill to Ramp Up SA Target to 75%," the *Guardian*, February 21, 2018, https://www.theguardian.com/australia-news/2018/feb/21/all-in-on-renewable-energy-jay-weatherill-to-ramp-up-sa-target-to-75.

12 "Growth and Low Carbon," Government of South Australia Department for Energy and Mining, 2021, https://www.energymining.sa.gov.au/growth_and_low_carbon.

13 J. Bassano, "SA Renewables Hit High Mark in 2020," *InDaily*, January 27, 2021, https://indaily.com.au/news/science-and-tech/2021/01/27/sa-renewables-peak-in-2020/.

14 Florian Zandt, "Infographic: Where Water Stress Will Be Highest by 2040," Statista Infographics, December 16, 2021, https://www.statista.com/chart/26140/water-stress-projections-global/.

15 "Markets and Overview 2021: Singapore Economic Forecast," Trading Economics, accessed October 30, 2022, https://tradingeconomics.com/singapore/forecast.

16 "Global Innovation Index 2021: Which Are the Most Innovative Countries," World Intellectual Property Organization, accessed October 30, 2022, https://www.wipo.int/global_innovation_index/en/2021/.

17 "Impact of Climate Change in Singapore," National Climate Change Secretariat, accessed October 30, 2022, https://www.nccs.gov.sg/singapores-climate-action/impact-of-climate-change-in-singapore/.

18 "Key Targets. Singapore Green Plan 2030," Green Plan, accessed October 30, 2022, thttps://www.greenplan.gov.sg/key-focus-areas/key-targets.

19 Asit K. Biswas, "How Singapore's Water Management Has Become a Global Model for How to Tackle Climate Crisis," *The Conversation*, November 24, 2021, https://theconversation.com/how-singapores-water-management-has-become-a-global-model-for-how-to-tackle-climate-crisis-162117.

20 "NEWater," PUB, Singapore's National Water Agency, accessed October 30, 2022, https://www.pub.gov.sg/watersupply/fournationaltaps/newater.

21 K. L. Ebi, J. J. Hess, and P. Watkiss, figure 8.1, in "Chapter 8: Health Risks and Costs of Climate Variability and Change," in *Injury Prevention and Environmental Health*, 3rd ed., October 27, 2017, https://www.ncbi.nlm.nih.gov/books/NBK525226/figure/ch8.sec2.fig1/.

22 "Climate Change: 2021 in 5 Numbers," World Bank Group, December 16, 2021, https://www.worldbank.org/en/news/feature/2021/12/16/2021-the-year-in-climate-in-5-numbers.

23 Florian Zandt, "Infographic: The World's Worst Offenders for Plastic Pollution," Statista Infographics, December 16, 2021, https://www.statista.com/chart/22959/metric-tonnes-of-plastic-packaging-produced-annually/.

24 "Visual Feature: Beat Plastic Pollution," UN Environment Programme, accessed October 30, 2022, https://www.unep.org/interactives/beat-plastic-pollution/.

25 David Matthews, "Are 26 Billionaires Worth More Than Half the Planet? The Debate, Explained," *Vox*, January 22, 2019, https://www.vox.com/future-perfect/2019/1/22/18192774/oxfam-inequality-report-2019-davos-wealth.

26 A. Chattopadhya et al., "Dilmah: Committed to Taste, Goodness, and Purpose," INSEAD Publishing, accessed October 30, 2022, https://publishing.insead.edu/case/dilmah-committed-taste-goodness-and-purpose.

27 "Dilmah Founder's Message," Dilmah, accessed October 30, 2022, https://www.dilmahtea.com/sustainability/dilmah-founders-message.html.

28 "What We Do: Dilmah Tea, Sri Lanka Luxury Tea Brands," Dilmah, accessed April 10, 2022, https://www.dilmahtea.com/dilmah-family/what-we-do.

29 "Our Core Values," Patagonia, accessed October 30, 2022, https://www.patagonia.com/core-values/.

30 Andrew Weaver, "Patagonia Rated Most Reputable Company in the U.S.," *Outside Business Journal*, May 17, 2021, https://www.outsidebusinessjournal.com/brands/camping-and-hiking/patagonia-rated-most-reputable-company-in-the-u-s/.

31 "Environmental Responsibility Programs," Patagonia, accessed October 30, 2022, https://www.patagonia.com/our-responsibility-programs.html.

32 D. Gelles, "Ryan Gellert, Patagonia's CEO, Has a Mission: 'Save Our Home Planet,'" *New York Times*, December 10, 2021, https://www-ny-times-com.cdn.ampproject.org/c/s/www.nytimes.com/2021/12/10/business/ryan-gellert-patagonia-corner-office.amp.html.

33 "Climate Goals," Patagonia, accessed October 30, 2022, https://www.patagonia.com/climate-goals/.

34 A. Engel, "Inside Patagonia's Operation to Keep Clothing Out of Landfills," the *Washington Post*, August 31, 2018, https://www.washingtonpost.com/business/inside-patagonias-operation-to-keep-you-from-buy-

ing-new-gear/2018/08/31/d3d1fab4-ac8c-11e8-b1da-ff7faa680710_story.
html.

35 R. Matthews, "10 Reasons Why Patagonia Is the World's Most Responsible Company," the Green Market Oracle, September 10, 2021, https://thegreen-marketoracle.com/2021/09/10/10-reasons-why-patagonia-is-worlds-most/.

36 Achim Berg et al., "Fashion on Climate," McKinsey, August 26, 2020, https://www.mckinsey.com/industries/retail/our-insights/fashion-on-climate.

37 P. Polman and A. Winston, *Net Positive: How Courageous Companies Thrive by Giving More Than They Take* (Cambridge, MA: Harvard Business Review Publishing, 2020).

38 M. Meisel et al., "Erfolgsversprechende Demand Response-Empfehlungen im Energieversorgungssystem 2020," *Informatik-Spektrum* 36 (2013): 17–26.

39 "Net Zero by 2050—Analysis," IEA, accessed October 30, 2022, https://www.iea.org/reports/net-zero-by-2050.

40 "Impact Report 2020," Tesla, accessed October 30, 2022, https://www.tesla.com/ns_videos/2020-tesla-impact-report.pdf.

41 M. Lewis, "Hertz Orders 100,000 Teslas, the Single-Largest EV Purchase Ever," Electrek, October 25, 2021, https://electrek.co/2021/10/25/hertz-orders-100000-teslas-the-single-largest-ev-purchase-ever/.

42 J. Gifford, "Oxford PV Completes 100 MW Factory Build Out," *PV Magazine International*, July 23, 2021, https://www.pv-magazine.com/2021/07/23/oxford-pv-completes-100-mw-factory-build-out/.

43 N. Wires, "Google Parent Alphabet Nearly Doubles Annual Profit," France 24, February 2, 2022, https://www.france24.com/en/business/20220202-google-parent-alphabet-nearly-doubles-annual-profit.

44 M. Bergen, "Google's CEO: 'We're Losing Time' in the Climate Fight," Bloomberg, October 17, 2021, https://www.bloomberg.com/news/features/2021-10-17/google-ceo-climate-fight-forces-us-to-push-boundaries.

45 "Our Commitment to Sustainability," Google, accessed October 30, 2022, https://sustainability.google/commitments-europe/#.

46 Wageningen University & Research, accessed October 30, 2022, https://www.wur.nl/en.htm.

47 "Precision Agriculture—Smart Farming," Wageningen University & Research, accessed October 30, 2022, https://www.wur.nl/en/Dossiers/file/dossier-precision-agriculture.htm.

48   F. Viviano, "How the Netherlands Feeds the World," *National Geographic*, 2021, https://www.nationalgeographic.com/magazine/article/holland-agriculture-sustainable-farming.

49   A. Boretti, S. Al-Zubaidy, M. Vaclavikova, et al., "Outlook for Graphene-Based Desalination Membranes," *NPJ Clean Water* 1, no. 5 (2018). https://doi.org/10.1038/s41545-018-0004-z.

50   Alberto Brambilla, "EU Can Cut Russian Energy Dependence Sooner Than Thought: Draghi," Bloomberg, April 17, 2022, https://www.bloomberg.com/news/articles/2022-04-17/eu-can-cut-russian-energy-dependence-sooner-than-thought-draghi EU 2022.

51   "A European Green Deal," European Commission, https://ec.europa.eu/info/strategy/priorities-2019-2024/european-green-deal_en.

52   A. Ng, "China's Pledge to End Building Coal Plants Abroad Improves Belt and Road's Reputation, Development Bank Says," CNBC, September 24, 2021, https://www.cnbc.com/2021/09/24/chinas-pledge-to-stop-building-coal-plants-abroad-helps-bri-aiib.html.

53   "China's Massive Belt and Road Initiative," Council on Foreign Relations, https://www.cfr.org/backgrounder/chinas-massive-belt-and-road-initiative.

54   Wang, "China Belt and Road Initiative (BRI) Investment Report 2021," Green Finance and Development Center, accessed October 30, 2022, https://greenfdc.org/brief-china-belt-and-road-initiative-bri-investment-report-2021/.

55   "Financing Climate Futures: Rethinking Infrastructure 2021," Organisation for Economic Co-operation and Development, https://www.oecd.org/environment/cc/climate-futures/.

56   *Banking on Climate Chaos 2021*, Rainforest Action Network, March 24, 2021, https://www.ran.org/wp-content/uploads/2021/03/Banking-on-Climate-Chaos-2021.pdf.

57   T. Espiner, "Big Banks Fund New Oil and Gas Despite Net Zero Pledges," BBC News, February 14, 2022, https://www.bbc.com/news/business-60366054.

58   "Werner Hoyer," European Investment Bank, https://www.eib.org/en/about/governance-and-structure/statutory-bodies/management-committee/members/werner-hoyer.htm.

59   Luke V. Wassenhove and Ravi Fernando, "Banking on EIB to Lead Sustainable Finance to Impact the Climate Emergency," INSEAD Publishing, April 23, 2021, https://publishing.insead.edu/case/banking-eib-lead-sustainable-finance-impact-climate-emergency.

60   Aman Kidwai, "The Outcome of COP26 Got Mixed Reviews. These 2 Business Leaders Left Feeling Hopeful," *Fortune*, November 15, 2021, https://fortune.com/2021/11/15/cop26-news-key-takeaways-deal-agreement-business-bcg-banco-santander/.

61   S. Pavoni, "Santander's Ana Botin on the Challenges of Sustainable Finance," FDI Intelligence, October 29, 2020, https://www.fdiintelligence.com/article/79020.

Chapter 5. Building a Twenty-First-Century Strategy Template

1   R. A. Fernando, Twenty-First-Century Board Leadership Model, US copyright TX 8-880-097, 2020.

Chapter 6. Are You Ready to Fight the Code Red for Business?

1   *WMO Atlas of Mortality and Economic Losses from Weather, Climate, and Water Extremes (1970–2019)*, World Meteorological Organization, https://library.wmo.int/index.php?lvl=notice_display&id=21930#.

2   "Financing Climate Futures: Rethinking Infrastructure 2021," Organisation for Economic Co-operation and Development, accessed October 30, 2022, https://www.oecd.org/environment/cc/climate-futures/.

3   R. A. Fernando, Twenty-First-Century Board Leadership Model, US copyright TX 8-880-097, 2020.

www.ingramcontent.com/pod-product-compliance
Lightning Source LLC
Chambersburg PA
CBHW021407210526
45463CB00001B/256